MW00698607

REBUILDING THE FAMOUS
FORD FLATHEAD
BY RON BISHOP

TAB Books
Division of McGraw-Hill, Inc.

New York San Francisco Washington, D.C. Auckland Bogotá
Caracas Lisbon London Madrid Mexico City Milan
Montreal New Delhi San Juan Singapore
Sydney Tokyo Toronto

25 26 27 28 29 C W W   C W W   0

Library of Congress Cataloging-in-Publication Data

Bishop, Ron.
    Rebuilding the famous Ford flathead.

    Includes index.
    1. Automobiles—Motors—Maintenance and repair.
2. Ford automobile. I. Title.
TL210.B513   629.2′504   79-25230
ISBN 0-8306-9965-1
ISBN 0-8306-2066-4 pbk.

Other TAB books by the author:

No. 2075 *Troubleshooting Old Cars*

Dedication:

I dedicate this book to my Pop, Ronald C. Bishop Sr., who shared my interest in old cars . . . from that first old '52 Ford flathead, he was there to lend a hand, advise, and lend a dollar or two when the going got rough. I'll miss ya, Pop.

Acknowledgments

In a work such as this, it is next to impossible to acknowledge and give credit to all the sources of help and information cited. Bits and pieces were gathered from an endless list of technical manuals as well as an equally large number of individuals. I would like to thank the Ford Motor Company without whom this book would have been impossible; Tom Hutchinson, La Puente, California—flathead mechanic par excellence—whose technical advice and guidance is reflected throughout the book, guidance and advice greatly appreciated; Bill Cannon, Monrovia, California, who is editor/publisher of the superb do-it-yourself publication *Skinned Knuckles-A Journal of Car Restoration*, written by and for antique auto enthusiasts. Bill's literary guidance was (and still is) invaluable. What he has taught me about automotive technical writing, will keep me forever in his debt. Special thanks for the chance in *Skinned Knuckles*, Bill, and for all your help; and to Charlotte Cannon, Bill's "personality-plus" wife; my pretty young wife Denise, who (sometimes) uncomplainingly puts up with my old car antics and writings and helps me with the secretarial chores. My dog "Dummie", a very good listener and friend. John Waugh, Lester Goetz, Norm Tillman, Sig Caswell, Henry Miller and all my other friends, who helped with their kind words of encouragement.

Contents

Introduction

Out in the weeds, in old barns, piled sky high in junk yards, at swap meets, under the hoods of some fine old Fords and even at the end of an anchor chain, you'll find Henry's finest.

From its birth on March 9th, 1932, the Ford L-head design V8 motor has been and still is a good, strong and dependable, fuel efficient power plant. Its popularity and dependability, from Mom's grocery-getter to the Bonneville salt flats, goes undisputed. It responds very well to hot rod "souping-up" techniques, making it very desirable to those wanting performance. Another factor that makes this motor popular is parts availability. In a nutshell, you can get parts (new and used) very easily throughout the country and in many cases, at less cost than what a newer car's parts would cost.

The Ford flathead motor is a relatively easy motor to work with. Its simple design makes it easy for the novice to do his own work and its relatively inexpensive upkeep and rebuild make it ideal for the restorer on a budget. Rebuilding one is not only a practical project, it's a fun one. When completed and back in your car, it will go anywhere and on little fuel; 20-25 mpg plus is not uncommon, even in the heavier cars.

Even though a person does his own work in his garage and follows the guidelines laid down by this book, he will still have to call upon the services of a professional machine shop. For example, boring a cylinder in a block to bring it back to like-new specifications is a rather simple task, if you have the tooling. It is very doubtful that the average reader of this book would have a $15,000 boring bar and mill table. Some work will have to be farmed out; it's a common and accepted practice.

Rebuilding the Famous Ford Flathead is a complete and thorough "how-to-do-it" book, encompassing the 22 years this popular motor was in production, which also includes the production span of Mercury, 1939-1953.

In the rebuilding of this motor (or any motor for that matter) you will have to follow closely factory recommended tolerances and dimensions, with good machine shop and mechanic procedures. In light of the technological advances made in the auto industry since this motor went out of production in 1953, this book was written incorporating the techniques, experience, and diagnostic testing afforded by today's technology. This book will tell the experienced and novice mechanic all they will need to know about and how to do a complete, thorough and dependable engine rebuild.

We'll begin this project with the assumption that you already have a motor to work with. We'll begin with basic engine specifications and major differences, by years of manufacture. Next, we'll start the actual tear-down of the motor and begin our inspection. From there we'll get into rebuilding and follow-through until Henry's finest is once again on the road.

There will be chapters for those of you not in need of a complete rebuild covering the rebuilding of all the bolt-on equipment such as the carburetor, distributor, fuel, and water pumps and so on. We'll also be looking at each engine system separately: cooling, fuel, ignition, electrical, oiling, reciprocating and valve train. All of the needed tables and data sheets have been compiled along with a mountain of pictures and drawings. From time to time, I have called upon the expertise of flathead builder par excellence, Tom Hutchinson, whom I'll talk about later. All in all, it promises to be very complete and thorough. Stay tuned . . . it should be fun.

Ron Bishop

Chapter 1

The Ford Flathead Through the Years

In 1932, the first flathead V8 featured a 90-degree block of cast iron alloy, with an L-head/valve design, solid lifters (sometimes referred to as push rods or cam followers) and a gear-driven camshaft (Fig. 1-1). The bore and stroke of 3.0625 × 3.750 gave a cubic inch displacement (CID) of 221 cubic inches. Compression ratio was 5.5:1, with peak horsepower of 65 @ 3400 rpm. The cylinder heads had 21 studs, with water pumps mounted at the front of each head; they used 18mm spark plugs. Intake and exhaust valve diameters were 1.54 inches with stems of .312 inch; no valve seat inserts were used. Connecting rods were 7 inch center to center, with full-floating (insert type) rod bearings on 2 inch diameter crankpins. It had a forged crank that was counter-balanced and three 2 inch diameter, poured main bearings. It weighed 525 lbs.

A change to aluminum heads in '33 raised the compression ratio to 6.3:1 with peak horsepower of 75 @ 3800 rpm. 1934 saw no major changes. However, a Stromberg 48, two-barrel carburetor was now used to replace the single-barrel Detroit Lubricator carb. This single change raised horsepower to 85 @ 3800 rpm. No major changes in '35 or '36, but toward the end of 1936, Ford was now using steel pistons instead of aluminum.

1937 and the Debut of the V8-85

1937 saw the forerunner of the modern line of flathead V8's. The aluminum heads were redesigned, moving water outlets to the center of the head and water pumps to the upper front of the block. A new combustion chamber shape and the use of domed type pistons reduced engine knock by giving a better quench area. This reduced compression to 6.12:1. The famous Stromberg 97 carb replaced the 48. A new forged crank featured enlarged mains of 2.4 inches and insert bearings to replace the old style poured babbitt type. Although there was no horsepower change, this motor became known as the V8-85. No changes in '38 other than to 14mm spark plugs.

With the introduction of the Mercury in 1939, the entire V8 was redesigned so that it could be used in both cars. Ford retained the 3.0625 bore, while Mercury went to 3.1875 for 239 cubic inches. Heads were changed to 24 studs to provide better sealing. Compression was 6.1:1 Ford and 6.3:1 for Mercury. New horsepower ratings were 85 @ 3800 for Ford and 95 @ 3600 for Mercury. 1940 and '41 saw no major changes. In '42, the heads were modified to raise compression to 6.2:1 for Ford and 6.4:1 for Mercury. Horsepower increased to 90 @ 3800 for Ford and 100 @ 3800 for Mercury. In 1946, '47 and '48, the Ford and Mercury motors were identical. Compression was 6.8:1 with 239 CID; horsepower was 100 @ 3800 rpm.

Final Design and 100 hp

Both the Ford and Mercury motors underwent major surgery in '49 and carried this final design until production career ended in 1953. A stamped steel or a cast iron bellhousing replaced the integral unit of the earlier motors. Twenty-four stud heads were still used, but the water outlets were moved to the front of the block. The crank was improved slightly as the counterweights were changed and rod bearings were now the non-floating, insert type. Mercury used a forged crank with a 4 inch stroke, while Ford used a cast nodular iron crank stroked at 3.750 inches. These modifications resulted in 255.4 CID for Mercury and 239.4 CID for Ford. Final horsepower ratings were 100 @ 3800 for Ford and 112 @ 3800 for Mercury. The distributor was completely redesigned and moved to the upper front of the engine. Driven off the camshaft by a spiral bevel gear, it was now a single breaker point unit, with full vacuum spark advance and no centrifugal weights.

In a nutshell, you have the evolution of the famous flathead. We'll talk about these changes more as we approach each year motor in our rebuild. There were many small changes throughout the years to these motors and I'll dwell upon them, too, when the

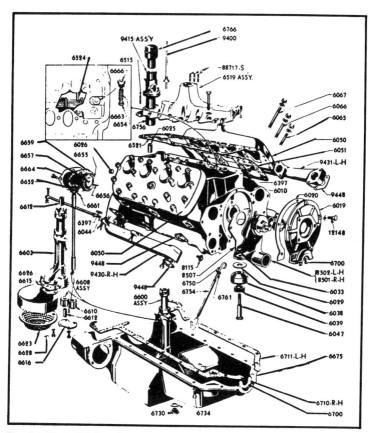

Fig. 1-1. Typical flathead motor.

time comes. Even though there are differences in these motors, many of the parts are interchangeable and the rebuilding procedures are fundamentally the same (Fig. 1-2).

Shopping for a Flathead

When shopping for a flathead motor, be selective; they're available and you have room to bargain. Here's a thumbnail guide as to what to pay for a motor, based upon my observations for the past year or so here in Southern California: Anything less is a bargain—anything more is a rip.

1933-1953: 21 and 24 Stud Blocks

☐ Complete, rebuilt and ready to install (with all accessories)— $750.

☐ Complete and in good running condition—$400.

☐ Complete and runs but needs to be rebuilt—$150.

☐ Complete, not running—$75.

☐ Short block with heads and oil pan—$50.

☐ Short block without heads—$25.

☐ Short block with crank only—$15.

☐ Bare block with main caps marked in place—$10.

☐ Bare block without main caps or a broken block—ZILCH. (New or used main caps and the machine work to align bore them will cost more than it's worth; unless it's a '32 block).

☐ Rebuilt short block—$500.

Prices could go higher or lower depending upon its overall condition. An exceptionally fine block or one with good working, new or rebuilt accessories will bring higher dollars. The key here,

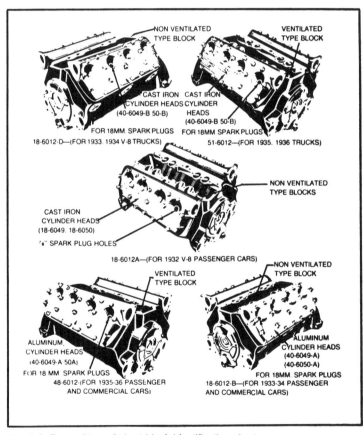

Fig. 1-2. Reconditioned short block identification chart.

is **be selective.** Relative to the rest of your car, a fine firing flathead should be one of your low dollar investments

1932 Only: The Rarest of the Flatheads

The guidelines above are out the window for '32 blocks. Here are approximate prices:

☐ Complete, rebuilt and ready to install (with all accessories)—$3000.

☐ Complete and in good running condition—$1500.

☐ Complete and runs but needs to be rebuilt—$1000.

☐ Complete, not running—$750.

☐ Short block with heads and oil pan—$500.

☐ Short block without heads—$350.

☐ Short block with crank only—$300.

☐ Bare block with main caps marked in place—$200.

☐ Bare block without main caps or a broken block—$75.

☐ Rebuilt short block—$1700.

Sources for Parts and Services

Ford Speed Parts Specialties
Tom Hutchinson
511 3rd Avenue
La Puente, California 91746
213/336-2128

Obsolete Ford Parts, Inc.*
6601 South Shields Boulevard
Oklahoma City, Oklahoma 73149
405/631-3933 or 405/631-7213
(Excellent source for drive line components)

Ford Parts Obsolete, Inc.*
1320 West Willow Street
Long Beach, California 90810
213/774-5460

Rick's Ford Parts*
P.O. Box 662
Shawnee Mission, Kansas 66201
Toll Free 1/800/255-4100
Catalog (168 pages) also free.

J. C. Whitney & Co. *
1917-19 Archer Avenue
P. O. Box 8410
Chicago, Illinois 60680
Catalog free.

Iskenderian *
16020 South Broadway
Gardena, California 90248
213/770-0930

Offenhauser *
5300 Alhambra Avenue
Los Angeles, California 90032
213/225-1307

Specialized Auto Parts, Inc. *
7130 Capitol Street
Houston, Texas 77011
713/928-3707
(Excellent source for drive line components)

Rock Valley Antique Auto Parts
122 South Pine
Stillman Valley, Illinois 61084
815/645-8168

Hemmings Motor News
Dept. 36, Box 100
Bennington, Vermont 05201
(A subscription to this fine publication can be of real help when trying to locate parts and services.)

*   Catalog Available
The above list of sources have always provided me with fast, reputable service. Although this list is small, it is not inclusive; check out the vendors in your area. New and used Ford flathead parts are readily available throughout the country.

Chapter 2
Initial Disassembly

With the assumption that everyone has a motor of some fashion to work with, we'll move out into the garage and get into the nitty gritty of this project. For the most part, all flatheads are basically the same and all procedures presented here will apply to all years of motors. When there is a difference, I'll dwell upon that difference as applicable.

Accessory Removal

If you have an engine stand, you are truly fortunate. If not, the garage floor and a few blocks of wood will suffice for now. Once you have removed as much crud as you can from the engine, the first logical step in the disassembly of your motor is the removal of the accessories. Some organization here would be prudent; gather some boxes, paper bags or whatever and separately, store each item with its hardware clearly marked as to content. We'll get to these items later in the series.

If you haven't done it yet, drain the oil from your motor and remove the following accessories:

☐ Carburetor and fuel line.

☐ Fuel pump and adapter. Lift out the fuel pump push rod from the intake manifold.

☐ Fan. Two types were used: the generator mounting bracket type and the crankshaft pulley mounting type.

☐ Generator.

☐ Spark plug wires, conduits and spark plugs.

☐ Distributor.

☐ Oil filter and oil lines.

☐ Starter.

☐ Water pumps. On the '37-'53 motors, one of the four bolts that holds the water pumps to the block is accessible only through the pump inlet opening. More likely than not, your only course of action is to drill the heads of these bolts off to remove the pumps. (Dumb place to have put a bolt.) We'll get the studs out of the block later on.

☐ Pressure plate and clutch disc: Relieve the tension on the pressure plate by pressing the three clutch release levers down and inserting a ⅜ inch thick wooden block between the lever and the pressure plate, before removing the hardware (Fig. 2-1).

☐ Flywheel. Lightly tap one side, then the other with a rubber mallet until the flywheel comes loose from the crankshaft. Remove the bolts first.

Engine Disassembly

With our accessories removed and temporarily stored away, we're now ready for the disassembly of the stripped block. This sequence includes the removal of the following:

☐ Oil pan. Hope you've drained the oil. If not, you're going to have a slick mess real soon.

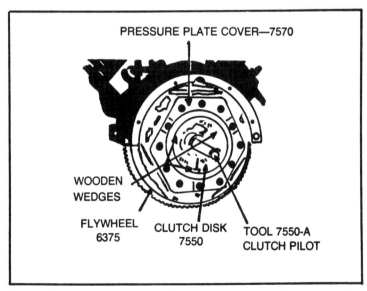

Fig. 2-1. Pressure plate, clutch and flywheel assembly. Note the location of the wooden blocks as mentioned in the text.

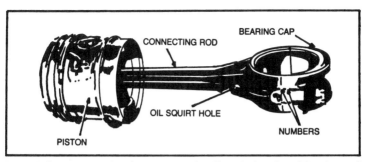

Fig. 2-2. Connecting rod and piston assembly. Note the location of the stamped numbers on both the rod and cap. The block adjacent to each rod should also have the same number stamped on the flange where the oil pan attaches. The main bearing caps should be marked in a similar fashion.

☐ Oil pump.

☐ Exhaust manifolds.

☐ Timing gear cover. Sometimes refered to as the cylinder front cover.

☐ Intake manifold.

☐ Cylinder heads. If your motor is old and rusty, you're going to have trouble here. Some of the nuts will come free, some won't, some will come out with the studs and others will break off. Get yourself a ½ inch drive socket and a good long breaker bar for plenty of leverage. Have someone hold the motor if it's not on a stand and fire away. Don't make the foolish mistake of trying to save the studs. They'll come out later very easily (promise) and they should be replaced anyway. More on this later. For now, just get the heads off.

☐ Connecting rods and piston assemblies (Fig. 2-2). If you plan to reuse your pistons, the ridge at the top of the cylinder bores must be removed. If not removed, the piston rings will grab on to these ridges as the pistons are forced through the bores and will break your rings and pistons. A special tool that will remove the cylinder ridge can be rented or purchased from Wards, Sears or the like for about $12.00. If you have a friend with one or you're in good with your local parts jobber, you can probably borrow one. It's not a tool that is used that often and need not be in your inventory.

You should own a set of ⅛ inch numeral metal stamps. A set running from 0-8 will run about $5.00 and will have many uses around your shop . . . here comes one now. Before we pull the con rod caps and push the rods and pistons out of the block, it would be wise to mark them as to location and position. From 1-8, mark each cap and rod on one side as to its corresponding cylinder. I like to use the motor's firing order, but any sequence is fine as long as you understand it. I mark the cylinder block flange that attaches to the oil pan. These marks will endure the hot tanks, stripper and machine shop.

There are several reasons for doing this: most all rods are rebuildable in these motors and if they pass magnaflux, they are worth saving; if they should become mixed, they will not have to be realign-bored; it keeps everything in its originally fit position now and in the future, on a balanced assembly, this is very important; and if you neither plan nor need to rebore, everything *must* go back together as originally installed, so as to fit its worn-in position.

Remove the nuts from the con rods and tap the con rod bolts with a brass, lead or plastic hammer until the con rod cap is free from the con rod. Get a small length of rubber tubing and make a shield to protect the con rod bolts. This will prevent any scoring to the bearings, crank or cylinder walls as the piston and con rod are pushed out of the cylinder block. An 18 inch piece of ½ inch dowel wood and your rubber mallet will handle this; the dowel is placed down through the block until it finds a resting stop on the underside of the piston face. Set the rod and piston aside and remove the con rod bearings. As each half is removed, clean it with kerosene or similar solution and mark it with a small piece of masking tape as to cylinder number, upper (con rod side) or lower (cap side). (If you plan on or need new bearings or crank work, you can bypass this operation). Put the bearing halves back into the con rods, reinstall the caps and nuts and set the assembly aside for now.

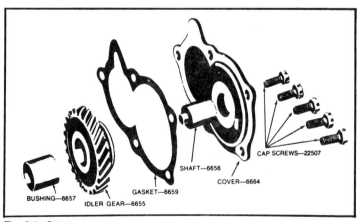

Fig. 2-3. Oil pump drive gear cover and related components.

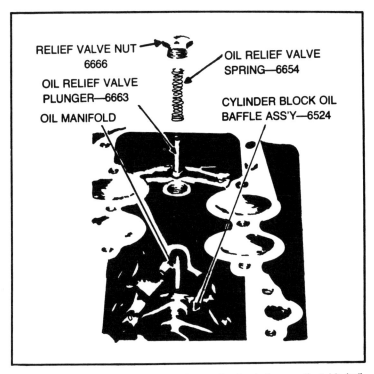

RELIEF VALVE NUT
6666

OIL RELIEF VALVE
PLUNGER—6663

OIL MANIFOLD

OIL RELIEF VALVE
SPRING—6654

CYLINDER BLOCK OIL
BAFFLE ASS'Y—6524

Fig. 2-4. The oil pressure relief valve is a spring-loaded, non-adjustable ball type, located under the intake manifold at the front of the motor. The flat spot on the ball is not wear, but a flow valve to allow a small amount of oil to flow when there isn't enough pressure in the system (usually at idle) to overcome the spring tension. The early V8's have a screw adjusted tension spring and the later V8's have their relief valve built into the oil pump. More on this later, when we get into the oiling system.

Remove the cap screws that hold the oil pump drive gear cover (Fig.2-3) in place and remove the oil pump drive cover, gasket and oil pump idler gear from the block. Set aside and store. If your engine does not have an oil relief valve built into the oil pump, it has one in the block. On these motors, remove the relief valve nut (Fig. 2-4), spring and plunger; set aside and store. Using a screw driver, pry under each oil baffle by prying its mounting clamp off the oil manifold; set aside and store.

We've accomplished a lot to get our motor to the "short block" stage and if you've come this along this far, you deserve a break (and so do I). We'll stop here and get ourselves a bit organized, clean up the mess we've made, and organize and store our parts temporarily.

Chapter 3
Short Block Disassembly

Before we get into the final part of our "tear down" and remove the cam, crank and valve gear, there are several tools that we are going to have to make, beg, borrow or steal and one we will have to buy. They're simple and inexpensive, so don't panic, but they will be of tremendous value and convenience during our rebuild.

Tools needed:

☐ Valve Spring Lifting Tool, (Fig. 3-1). This is the only item we will have to buy. They are readily available from antique Ford houses and you may find that a well-stocked tool supply store will still carry them. Rick's Antique Ford Parts, Box 662, Shawnee Mission, Kansas 66201, sells this tool for $5.95; (Tool #8520-0, Page 4 of Catalog #316). Rick has a toll free number: 1-800-225-4100.

☐Crankshaft Support, (Figs. 3-2 and 3-3). There are two ways to properly store a crankshaft when it is out of a motor; vertically, which can become a nuisance and very easy to knock over; and horizontally, which is the best way, but if not done properly, you could very well ruin your crank. The crank must be supported by its main bearing journals or it will sag under its own weight, over a long period of time. It won't do contortions, but it could very possibly sag .015" to .020" or more. Once this happens your crank is worthless; they cannot be straightened economically. To avoid this problem and to provide yourself with a convenient working and inspection vise, take the time to construct a small storage bench

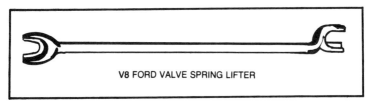

V8 FORD VALVE SPRING LIFTER

Fig. 3-1. V8 Ford valve spring lifter.

out of some wood you probably already have laying around the garage.

☐ Camshaft Support Block: Everything I just said about the crank holds true here also . . . make a bench.

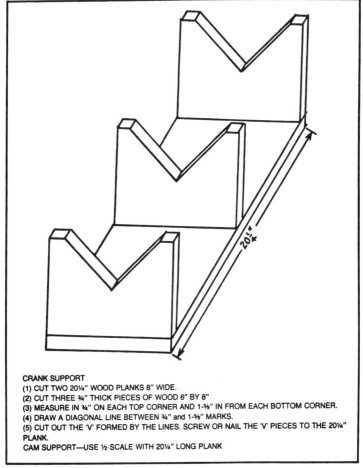

CRANK SUPPORT
(1) CUT TWO 20¼" WOOD PLANKS 8" WIDE.
(2) CUT THREE ¾" THICK PIECES OF WOOD 6" BY 8"
(3) MEASURE IN ¾" ON EACH TOP CORNER AND 1-⅝" IN FROM EACH BOTTOM CORNER.
(4) DRAW A DIAGONAL LINE BETWEEN ¾" and 1-⅝" MARKS.
(5) CUT OUT THE 'V' FORMED BY THE LINES. SCREW OR NAIL THE 'V' PIECES TO THE 20¼" PLANK.
CAM SUPPORT—USE ½·SCALE WITH 20¼" LONG PLANK

Fig. 3-2. Crank and camshaft support.

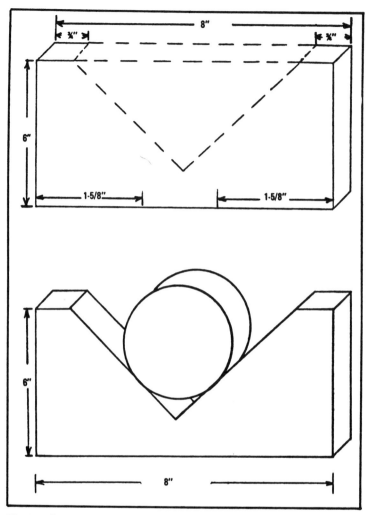

Fig. 3-3. Crank and/or cam support block.

☐ Valve Rack, (Fig. 3-4). Sixteen valves can become mixed very easily but they all have to go back into the same holes they came out of. To keep track of them easily, get a board about 14 inches long and drill 16 somewhat evenly spaced 1-⅛ inch holes in it; number the holes as to cylinder and if it is an intake or an exhaust: #1E and #1I, #2E and #2I, etc.

☐ Hook Tool, (Fig.3-5). The valve guide bushing retainer (the horseshoe shaped critter) can be removed easily when the valve spring is compressed, with the use of a special tool. Take a

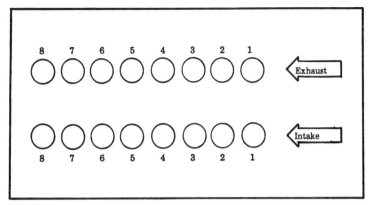

Fig. 3-4. Valve rack.

heavy piece of welding rod about 18 inches long and bend a hook on one end of it. Hook the tool in the retainer, compress the valve spring and pull the retainer upward. When the retainer is out, the valve assembly should lift out of the block . . . nothing to it.

Now that we have some tooling made, we can get to the final stage of this "tear down" and send the block and a few other parts to the stripper or the hot tanks.

Crankshaft Removal

Again, we'll use our metal stamps and mark the main bearing caps before their removal. Stamp the cap and main web on one side only: #1—Front, #2—Center and #3—Rear. Failure to do this could result in an unneeded and somewhat expensive align boring. In all but a few motors produced prior to the Model 78 in 1937, the crankshaft is supported by three babbitt bearings which must be poured and aligned bored. (Fig. 3-6.) (Definitely a job for experts.) With a little care and some precaution, insert-type bearings used in all motors from '37 on will not need their main caps realign bored, provided of course there are no misalignment problems with the block. Your machine shop will automatically check this when the block is in for boring and camshaft bearings.

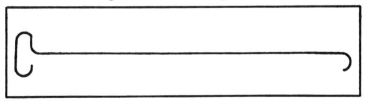

Fig. 3-5. Valve keeper hook tool.

The bearings should be cleaned and marked (like the connecting rod bearings) as they are removed from the engine. Remove the hardware (cotter pins and nuts) from each main bearing. Place a screwdriver between the cylinder block and the boss provided on each side of the front, center and rear bearing cap and pry the caps to release them from the cylinder block. Lightly tapping the caps with a plastic hammer while prying will help ease them off. Lift the bearing caps off the crankshaft along with the lower main bearing (don't forget to clean and identify it). Lift the crank from the block and place it in the wooden support tool we made earlier and remove, clean and mark the upper main bearings. Set all six main bearings aside and store them in a safe place. Now replace the main bearing caps in their original position by matching the metal stamped numbers. Install the nuts and snug them down.

Valve Assembly Removal

The valve, split guide, spring and retainer are installed and removed as an assembly, which are locked in place by the horseshoe shaped guide retainer. With the valve in the closed position (rotate the camshaft if necessary) engage the end of the special tool in the notches in the lower end of the valve guide and pull the guide down so that the guide retainer can be pulled out with the hook tool we made earlier (Fig. 3-7). Lift the valve assembly and lifter out of the block (Fig. 3-8.) No need to disassemble the valve assembly now, but whenever handling the split guide, keep the halves paired properly as the inside and outside diameters are related and held to close tolerances. Some masking tape and your felt tip pen will do the job. In the same manner, so identify the valves and the springs and lifters as to the cylinders and valve port from which they were removed. For now, store them in the valve rack we made earlier.

If the motor is old and has not been in service for a long while, the valve guides may be so tight in the block that they cannot be pulled down with the lifter. When this happens, use the lifter to compress the valve spring and remove the spring retainer. Lift the valve as high as it will go in the guide and drive the two halves of the guide downward by inserting a thin, blunt drift down through the valve opening. The guide retainer can now be removed and the valve and guide assembly pried out with the lifter bar.If this doesn't work and the valves are still stuck, you might try a little heat. If heat doesn't work, you're going to have to waste the valve and cut the lower end of it off in order to get it out of the block. If your

24

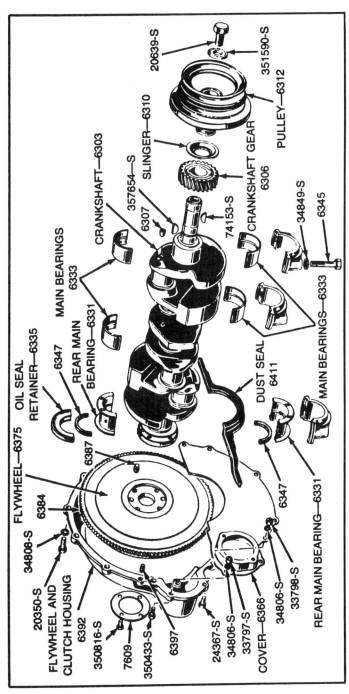

Fig. 3-6. Typical crankshaft and related components.

25

valves are in this condition, you're really not wasting anything, as the valves are junk anyway. To remove them the "hard way" fire up your cutting torch and blast away at the retainer end of the valve until everything comes unglued. This is a bit "Mickey Mouse," but effective. The only other alternative to this is to send your block to the hot tanks or to the stripper with the valves and camshaft intact. The disadvantage to this is that you will have to send the block twice to get it cleaned once. If you opt for this method and you have babbitt bearings, check with your stripper first, or at least tell him you have babbitt cam bearings in your motor . Babbitt will dilute some of the chemicals used in commercial stripping and if you ruin the man's tanks, you may find yourself with a lawsuit on your hands. At the very least, you will have made an arch enemy.

Camshaft Removal

The cam rotates in three replaceable, babbitt-lined bearings, with replacement best handled by a machine shop (the tools required are too expensive for us to handle at this garage level rebuild.) Two types of camshafts have been used in flatheads. The earlier models have a pressed-on drive gear and the later a bolt-on gear. All camshafts are interchangeable from motor to motor. The cam should slide out of the block with no hassles if it is a free turning unit. If it is stuck, send the block to the hot tanks. Once the cam is out of the block, store it in your wooden storage block.

Off to the Stripper

At this stage of the game, your block is about as bare as old Mother Hubbard's cupboard, and ready for the stripper. A few old gaskets, seals and some frozen studs may still be hanging on, which is of no consequence. Go ahead and send your block to the stripper (don't forget about the babbitt cam bearings and/or main bearings) along with any other parts that may need a good cleaning. The oil pan front timing cover, rear oil cover, the oil pump, exhaust manifold, intake manifold and the heads are good candidates. Don't send your crank, cam, pistons, rods or valve gear at this time, or any of the accessories.

Preliminary inspection of our block will be limited to a visual check of the main webbing support bosses. If you can see a crack or one of them is broken, sorry, but your motor (at least the block) is junk. Upon return of the block from the stripper, we'll take it to a machine shop and have it "magnafluxed." If we pass mag, we're home free.

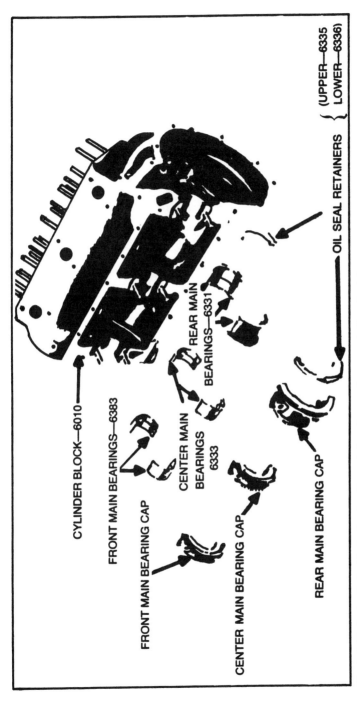

CYLINDER BLOCK—6010

FRONT MAIN BEARINGS—6383

FRONT MAIN BEARING CAP

CENTER MAIN BEARINGS 6333

CENTER MAIN BEARING CAP

REAR MAIN BEARINGS—6331

REAR MAIN BEARING CAP

OIL SEAL RETAINERS { (UPPER—6335 LOWER—6336)

Fig. 3-7. Typical cylinder block showing the main bearings and caps and oil seal retainers.

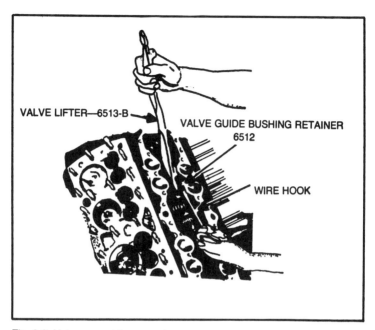

VALVE LIFTER—6513-B

VALVE GUIDE BUSHING RETAINER 6512

WIRE HOOK

Fig. 3-8. Valve assembly removal.

It may seem like a lot of unneccessary time and money is being spent to clean and inspect a block that we may not be sure is good enough to be used, but there's no other way. You can't properly inspect or mag a dirty block, and you can't use the block unless it's been inspected (you can, but you would be very foolish to do so). Take heart though, you're rebuilding a Ford Flathead and not a flimsy overhead. These blocks are probably the strongest and castings ever made or used in American automobiles. They are also one of the cleanest castings ever produced; notice how little casting flash there is throughout the block. That's a sign of quality construction. These motors are strong and were designed and built to last . . . over-engineered if you will. Chances are about 80 percent in favor of having a good block if you can't visually detect anything wrong. The magnaflux is cheap insurance should there be anything wrong. It would be better to find out now that the block is no good than later when you have about $150 or so of machine work into it.

Chapter 4
Inspection and Machining

We're going to do a thorough inspection of our block, reciprocating assemblies and support items. We'll get any broken, damaged or frozen studs out of the block and we'll also determine what will have to go to the machine shop and what will and should be done, once it's there. There will be a lot of material to absorb and digest this go-round, so before we get started take care of your domestic chores and tell the wife you'll be out on the patio with a "cool-one," 'cause ya got some engineering to do.

In the last chapter, we sent our block to the stripper along with a few other items. We should have the block back by now, so we can start our inspection. You'll notice that the few extra dollars spent to send the block to the stripper rather than to the hot tanks was well worth it. The block is immaculate and you can probably eat off those water jackets. There's no way could the hot tanks get that scale out of them. Everything and anything we can do to help keep this motor cool will be to our advantage, once we're on the road again. Clean, unrestricted water jackets are essential.

The first thing to do is mount the block on an engine stand or get two men and a small boy to help heft it up to the work bench. Inspect the main webbing support bosses. Any cracks or broken pieces is cause enough to go searching for another block. There's no way can they be repaired . . . sorry. Small breaks and chips at the bottom of the cylinder bores are of no concern, unless they are big (i.e., a ¼ inch or more in circumference). Check the cylinder decks for cracks. No doubt you'll have one or two cracks running from the stud holes to the water jackets. They're common and are nothing to worry about. If you have cracks running from the cylinder to a valve seat, a water jacket or a stud hole, you're going to have to have them welded or find another block.

If your motor was in running condition prior to disassembly and if the valve springs were rusted or if there was an excessive amount of sludge in the valve chamber, it is an indication that the block might be cracked and should be inspected thoroughly. Cracks between the valve seat and the cylinder bore are caused (aggravated) by poor water circulation and engine overheating. Any cracks from around the bell housing alignment dowel pins should also be repaired.

Cylinder Head Stud Removal

As promised earlier, we're going to get those broken and/or frozen studs out of that block. If you sent your block to the hot tanks, you're still going to have a problem. On the other hand, if you sent your block to the stripper, as I suggested, you've got 50 percent of the problem in back of you. Those hours the block spent in the stripping tanks will attack crud like you've never seen before.

At any rate, if you still have studs in the block that are not broken off, clean up the threads with a die and jam two nuts together on the stud. Tighten the stud just a bit; you'll probably hear a "cracking" sound. Then proceed with backing the stud out of the block. If your threads are too far gone for a die clean-up, try a pair of vise grips, using the same procedure just mentioned.

If the stud still won't come out or it breaks, proceed as follows. Soak, no, drown all studs with penetrating oil, WD40 or similar solution. Give the agents time enough to work, 20 minutes should be sufficient. Center punch the stud for drilling. Drill a ⅛ inch or smaller pilot hole in the stud about ¾ of the way down. Change your drill bit to a 5/32 inch and drill the pilot hole out. Install a #3 Easy-Out and try to back the stud out. If this fails or the easy out looses its traction, pilot the hole with 3/16 inch drill. Drill that pilot hole with a ¼ inch drill and install a #4 Easy-Out. Once all the studs are out of the block, run a 7/16 inch × 14 tap down the threads to clean them up.

Do this several times, being sure to use plenty of cutting oil, removing the cutting chips from the hole. If all the above fail, your local machine shop will have to drill the stud out and install a

Heli-Coil to save the stud hole and the block. Other broken studs throughout the block will be handled in the same fashion. Just make the appropriate drill, Easy-Out and tap changes.

If you're a bit gutsy, or familiar with these blocks and torch equipment, you can blast the studs out with a cutting torch. The block castings are strong enough to withstand this kind of punishment, but don't quote me on this procedure—it requires a bit of experience.

Fuel Pump Push Rod Bushing

The fuel pump push rod bushing should be replaced or at least checked with a new push rod for clearance and fit. This is not critical, but wear of .020 inch or more will allow oil to escape from the main oil passage, making it very difficult to build up and/or maintain proper oil pressure. Drive the old bushing from the block and install a new one. Use a piece of wood to absorb the shock and drive the new bushing in place until it is flush with the casting.

Deck Alignment

The cylinder block decks are those surfaces which contain the cylinders and valves, the surface to which the cylinder heads are bolted. Each deck should be flat, parallel and exactly the same distance from the main bearing bores. This is an area that was ignored 25 and 30 years ago when rebuilding an engine and money spent on this operation will be money well spent. Your heads will seal to the surface better and your cylinders will be balanced as to volume. In short, it will make a big difference in the way the motor performs. The block is secured in a saddle and squared to the mill table. Each deck is then cut with a large cutting wheel to square the deck flat and true and to equalize their distance from the crankshaft. Any competent machine shop can perform this operation.

Cylinder Bores

It will be necessary to rebore any cylinders:

☐ That have a taper of more than .006 inches.
☐ That are out of round by more than .003 inches.
☐ If wear exceeds .005 inches.

You must have the pistons you are going to use prior to boring your block. You can have your block bored to fit pistons, but it could be a problem to find pistons to fit a particular bore. Once you have the measurements of your cylinders, locate pistons that are enough oversize to allow for a clean up bore. Usually a .030 inch bore will be enough to clean up the cylinders. The machine shop will bore cylinders .0015″ under the size of the replacement pistons you bring in. The .0015″ undersize will be cleaned up to final piston size when the cylinders are honed. The pistons will be kept with the block and identified as to the cylinders they have been fitted to. (Usually by using a metal-stamped number corresponding to cylinder number on the skirt or face of the piston.)

Checking Cylinder Bore Sizes

Several expensive, precision pieces of equipment (micrometer and telescope gauge) will be required to accurately determine cylinder taper, roundness, wear and bore. Most machine shops will provide this service for free, but for those of you who have the equipment, here's how it's done:

☐ **Taper.** Measure the diameter of the cylinder lengthwise of the block at the bottom of the cylinder and at the deepest point of the ring wear (Fig. 4-1). Measure as before but this time crosswise to the block (90 degrees opposite). A comparison of the readings taken lengthwise, when compared to the readings taken crosswise, will indicate "taper."

☐ **Roundness.** Measure the diameter of the cylinder at the deepest point of ring wear lengthwise and crosswise. Now, measure the diameter of the cylinder lengthwise and crosswise at the bottom of the cylinder. A comparison of the readings taken at the deepest point of ring wear, when compared to the reading taken at the bottom of the cylinder, will indicate the "roundness" or "out-of-round" of the cylinder.

☐ **Wear.** A comparison of measurements taken lengthwise at the deepest point of ring wear, when compared to measurements taken lengthwise at the bottom of the cylinder, will indicate wear.

☐ **Bore.** A lengthwise measurement about 1 inch from the bottom of the cylinder will indicate the cylinders current bore.

Cylinder Sleeves

Cylinder sleeves have four useful purposes:

☐ They can be used to eliminate the need for boring when the cylinder wear is beyond limits. (Some flathead motors built after 1939 were so equipped).

☐ They can be used to repair a damaged or cracked cylinder.

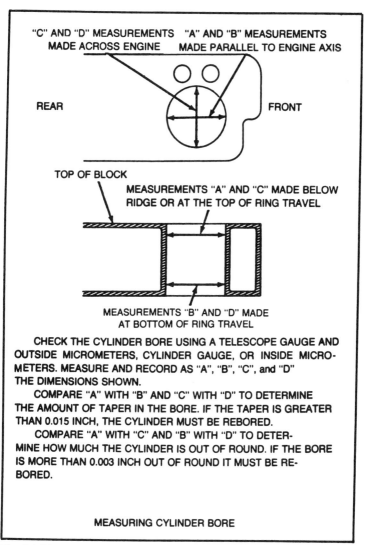

CHECK THE CYLINDER BORE USING A TELESCOPE GAUGE AND OUTSIDE MICROMETERS, CYLINDER GAUGE, OR INSIDE MICROMETERS. MEASURE AND RECORD AS "A", "B", "C", and "D" THE DIMENSIONS SHOWN.

COMPARE "A" WITH "B" AND "C" WITH "D" TO DETERMINE THE AMOUNT OF TAPER IN THE BORE. IF THE TAPER IS GREATER THAN 0.015 INCH, THE CYLINDER MUST BE REBORED.

COMPARE "A" WITH "C" AND "B" WITH "D" TO DETERMINE HOW MUCH THE CYLINDER IS OUT OF ROUND. IF THE BORE IS MORE THAN 0.003 INCH OUT OF ROUND IT MUST BE REBORED.

MEASURING CYLINDER BORE

Fig. 4-1. Measuring cylinder bore.

☐ They can be used to "hold" a cylinder to a particular size when a particular size piston must be used. Usually this is done when the engine's original cubic inch displacement (CID) is to be retained and/or the pistons that come from the block are in good repair (Table 4-1).

☐ They can be used to "size down" a cylinder that has been previously bored to its limit.

33

Table 4-1. Cubic Inch Displacement Chart.

Bore in Inches	3-¾" stroke CID	4" stroke CID	4-⅛" stroke CID
3-1/16	220	236	243
3-3/16	239	255	263
3-5/16	258	274	284
3-5/16 + 0.030	263	281	289
3-⅜	268	286	296

In all cases, sleeves are considered an economical and worthwhile repair. New, cast iron sleeves are manufactured with a finished cylinder bore to fit a standard size piston and do not require honing. They will provide the same service and dependability as a cylinder without a sleeve.

Sleeve installation is accomplished by reboring the cylinder .0012 inches undersize of the sleeve to be used, to facilitate the correct press fit. The cylinder is also counterbored at deck level (top of block) to accommodate the flange at the top of the sleeve. The sleeve is then pressed or drawn into the cylinder (Fig. 4-2).

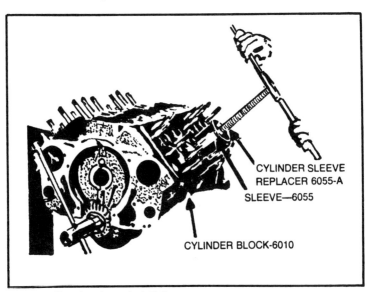

CYLINDER SLEEVE
REPLACER 6055-A

SLEEVE—6055

CYLINDER BLOCK-6010

Fig. 4-2. Cylinder sleeve installation.

Sleeve removal is accomplished by either pressing the old one out or by driving a pointed tool between the cylinder and sleeve, which will collapse the sleeve (Fig. 4-3). It can then be pulled out with a pair of pliers or vise grips.

Cylinder Honing and Piston Fitting

Rebore each cylinder .0015 inches undersize, required for the oversize piston to be used, leaving the .0015 inches for honing. A #220-L grit hone is commonly used. The honing operation requires the removing of just enough material to obtain the correct clearance for the pistons to be used. When fitting pistons, clean the cylinder walls and pistons thoroughly. Use a tension scale and thickness (feeler) gauge to obtain the proper "fit." Pistons must be kept with the cylinder block and each piston must be marked indicating the cylinder number to which it has been fitted.

To check the clearance of a piston in a cylinder bore, use a feeler gauge ½ inch wide and long enough to cover the entire length of the piston. Attach the gauge to a tension scale. Place the gauge on the side of the piston bore and push the piston into the cylinder, so that the gauge is between the piston and the cylinder bore. Withdraw the gauge and record the reading on the tension scale. The thickness of the gauge to be used and the pounds pull for the various combinations of pistons and cylinder bores are as follows:

Table 4-2. Bore and Piston Combinations.

Bore & Piston Combinations	Steel & Aluminum Pistons	
	Gauge Thickness	Pounds Pull
New sleeve and new piston	.003	6-10
Worn sleeve and new piston	.004	6-10
Worn sleeve and worn piston	.005	6-10
New bore and new piston	.0025	6-10
Worn bore and new piston	.004	6-10
Worn bore and worn piston	.005	6-10

Ford engines used both steel and aluminum alloy pistons, according to the production year and application. It is possible to interchange them as long as dome-head pistons and flat-head pistons are used with the proper heads. With dome-head pistons, use the cylinder head having a combustion chamber which covers the entire cylinder bore; with flat-head pistons, use a head having the smaller combustion chamber which does not cover the entire cylinder bore. Don't mix any two piston styles in motor. The diameter measurement of split-skirt type pistons must be made at the top of the skirt, just under the oil ring groove and at a right angle to the piston pin. Trunk-type piston diameter is measured at the lower end of the skirt and also at a right angle to the piston pin.

Proper honing is a must for proper ring sealing and longevity. Use a 220 grit hone, lubed with 10W or 20W oil only. The crosshatch pattern made by the hone should be between 30 degrees and 45 degrees from the horizontal. This is regulated by the speed at which the hone is moved up and down in the cylinder. After honing, wipe the cylinder with an oil saturated rag (start with a clean rag) to float the grit to the top of the grooves left by the honing. Then use solvent or lacquer thinner to remove the oil and any leftover residue. Don't use the solvent or thinner without first using the oil. Solvent or thinner won't effectively remove the honing grit, because it will not lift the particles out of the grooves.

Valve Seats

A valve seat insert will have to be replaced if:
- ☐ It is cracked.
- ☐ It is loose within the cylinder block.
- ☐ The width of the seat measures .125 inches or more.
- ☐ New guides are to be installed.

Valve seat insert replacement and refacing is best handled by a professional machine shop. Here's what will take place. The valve seat will be pulled from the block and the counterbore machined to obtain a .0015 to .0030-inch press fit on the replacement insert. The insert will be packed in dry ice for 10-15 minutes to shrink the seat slightly, then driven into the block with a mandrel guide tool. As the insert warms to room temperature, it will expand slightly and snug itself in the valve insert bore. The seat will then require a 90 degree valve angle machined into it. The width of the valve seat should not be more than .125 inch after refacing, if measured across the face of the seat. Proper width will fall within .062 inch.

The machine shop will grind what is known as a "triple valve seat". This consists of a 90 degree grinding wheel that will make the first cut on the valve seat, 120 degree cutter for removing material from the top of the seat and a 60 degree cutter for removing material from the bottom of the seat. If the grinder and pilot are in good condition, hand lapping of the valves will not be needed in order to insure a good valve seal.

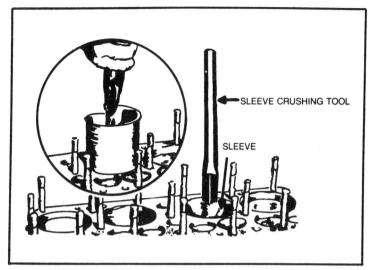

Fig. 4-3. Cylinder sleeve removal.

A valve is lapped by using a special abrasive compound on the seat of the valve. The valve is then rapidly rotated back and forth against the seat with the use of a special tool that attaches a small suction cup to hold and rotate the valve. Never match a precision ground valve to a lapped seat or vice versa; restrict valve lapping to valves that are in nearly perfect condition. A bit of advice . . . insist upon precision grinding of your valve seats. Lapping a valve to its seat is inferior and archaic. Also, the valves and valve seats should be ground when you have the block in for other machine work, rather than after the block has been cleaned for final assembly.

A 45-degree valve angle on the seat and .060-inch seat width is what a stock motor should have when the valve seat grinding machining is completed. However, you can gain a small bit of performance (fuel flow) by putting a 30-degree angle with a .090-inch width on the intake valve inserts, leaving the exhaust at 45 degrees by .060 inch.

Valves

The valves, when ready to be installed in the motor, will have been ground by the machine shop and matched port for port, valve seat to valve seat; 45 degrees intake and exhaust (stock); or, 45 degrees exhaust, and 30 degrees, intake, (performance). You can have your valves refaced if they are pitted, corroded, or burned provided they are not warped and will clean up with a light (.005

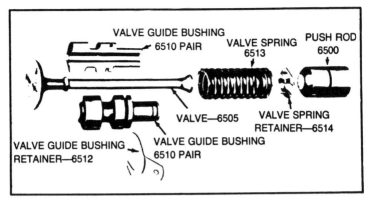

Fig. 4-4. Typical valve assembly.

inch) cut of the grinding wheel. If a cutter was used to reface or face your valve seats rather than a grinder, the valves will have to be lapped into the seats for a good valve seal.

Replace valves that are broken or that have bent or scored stems. Replace any valve that has stem wear less than .3065 inch. Valve stems that measure .3090 inch or more are satisfactory for use as intake valves; valve stems that measure .3065 inch or more are satisfactory for use as intake valves. Hard-chroming may possibly add .002-.003 inch. Stock and/or new valve stem diameters are .3115 inch for '32-'48 motors and .3410 inch for '49-'53 motors (Fig. 4-4). Valves and guides, as a unit, are interchangeable from year to year. The springs will be matched for length and pressure (Fig. 4-5.)

Pistons and Connecting Rod Disassembly

Once the pistons and con rod assemblies have been removed from the block, the caps reinstalled and the bearings marked, remove the two piston pin retainers and push the piston pin out of the piston. (These are not a press fit.) Reinstall the pin and retainers in the piston. Remove the four piston rings with a piston ring expander. Clean as much carbon from the top of the piston and the ring grooves as possible. Clean the entire assembly in kerosene. Discard any piston which is cracked, scored, has burn spots or is in any other way damaged.

Piston Pins and Bore

Replace piston pins that measure less than .749 inch. You can use a new piston pin as a gauge and insert it in the piston. If the pin falls through under its own weight, the pin bore in the piston is excessively worn and must be replaced with a new piston. The alternative to new pistons would be to have the pin bores reamed and burnished or honed to accommodate an oversize piston pin. For the money and hassles involved, spend a few extra dollars and go with new pistons and pins, which will already be fit to tolerance. Original piston pin diameter should measure from .7501 inch to .7504 inch.

If a connecting rod bushing or a piston pin hole is worn and the inside diameter does not measure more than .7535 inch, it can be honed or reamed and burnished to fit a .001 inch or .002 inch oversize piston pin. As a rule of thumb, the fit is correct for a piston pin in the connecting rod bushing if the pin will pass slowly through the bushing by its own weight. The fit for a piston pin in the piston is when it can be inserted in the piston under slight pressure with the piston at room temperature—about 70 degrees. When replacing the bushings in the con rod, the machine shop will drive the bushing from the con rod and press a new bushing in. The machine shop will then drill the four oil holes in the bushing to the same size as the holes in the con rod. The bushing will then be reamed and burnished (or honed) to .7505 inch. The rod will then be checked for any misalignment.

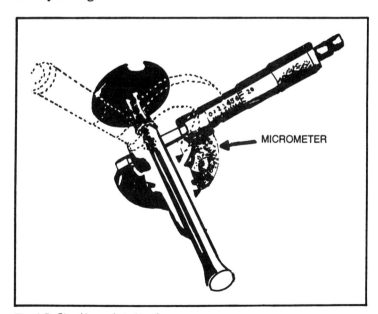

MICROMETER

Fig. 4-5. Checking valve stem for wear.

Final note on piston pins; new lock rings should always be used when the existing ones are removed for any reason. Also, a 1/32 inch clearance must be maintained between the ends of the piston pin and the lock ring.

Connecting Rods and Bearings

If the con rod crank pin bore is worn .0015 inch or more, it must be bored to accommodate the next size replacement bearing. If the bearing on one con rod is changed, it will be necessary to change the size of the other rod used on the same crankpin so that both may be used on the same size journal. As a rule, measure all eight rods and determine which will have to be bored the largest, then rework all the rods to that next over-size replacement bushing.

Replace con rod bearings that are worn, pitted, scored, or discolored (usually caused by overheating and/or not enough oil), or bearings that measure less than tolerance:

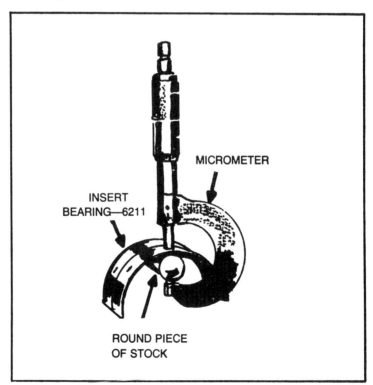

Fig. 4-6. Checking connecting rod bearing for wear.

☐ Stock Con Rod Bore Diameters. 1932-1942 2.220 inch, 1946-1948 2.298 inch, and 1949-1953 2.298 inch.

☐ Connecting rod bearings 1932-1948. These motors use a "free floating" CR bearing. The halves of these type bearings turn freely in both the rod and on the crank journal. There is no practical way to measure the clearance of these bearings when installed, therefore, the crankpin, rod bore and bearing shells must be measured separately (Fig. 4-6). Deduct the thickness of the round stock from the total measurement to obtain the bearing thickness. Replace standard bearings that measure less than .1085 inch and bore rods that measure more than 2.2215 inch.

☐ Connecting rod bearings 1949-1953. These motors use an insert-type bearing. The halves of these bearings are locked into place on the rod via a small key pressed into the shell of the bearing, which is held in place by a matching slot in the con rod. Like the earlier type bearings, there is no way to measure these bearings when they are installed. Therefore, the crankpin, rod bore and bearing shells must be measured separately. Deduct the thickness of the round stock from the total measurement to obtain the bearing thickness. Replace standard bearings that measure less than 0.1085 inch and bore rods that measure more than 2.2995 inch.

Crankshaft

Thoroughly clean the crankshaft with kerosene or similar solvent. Clean out all the drilled holes in the journals with a rifle brush or a piece of wire. There are four sludge traps in the crank journals that must be cleaned. Remove and replace the welch plugs. Replace the crank flange dowel pin if it is damaged. Replace the crankshaft drive gear that has chipped, broken, or worn teeth. Remove the gear with a gear puller that will pull the gear evenly. Remove the crankshaft gear Woodruff key.

Table 4-3. Original Crankshaft Journal Diameters.

Original Crankshaft Journal Diameters:		
Year	Con Rod	Main Bearing
1932-36	1.998-1.999	1.998-1.999
1937-38	1.998-1.999	2.398-2.399
1939-42	1.998-1.999	2.498-2.499
1946-49	2.138-2.139	2.498-2.499

To install the crank gear, tap the crank gear Woodruff key into the crank and press the gear onto the crankshaft. If the main or crankpin journals are grooved or scored, the crank will have to be remachined. Measure each journal diameter in at least four places to determine size, out-of-round and taper. Remachine a crank that:

☐ Has a journal out-of-round more than .0015 inch.
☐ Has a taper of more than .001 inch.

The maximum bearing clearance on a worn journal is .003 inch for main bearings and .005 inch for connecting rod bearings. Subtract the amount of undersize of the bearings to be used from the original size of the crankshaft and remachine the crankshaft by grinding to the new size. Polish with grit polishing paper, removing not more than .0009 inch from the diameter.

Ford used a cast nodular iron crank with a 3¾ inch stroke from '46-'53. All Mercurys used a forged crank with a 4 inch stroke that will drop into any '46-'53 block, without any modifications. This change has the advantage of a forged crank and an additional ¼ inch stroke over a stock Ford unit. You can tell the Ford crank from the Merc by tapping them with a metal hammer. The forged crank will have a much clearer ring to it than the cast unit, unless of course the forged unit is cracked.

The Merc crank can be stroked an additional ⅛ inch to 4-⅛" by grinding the rod journals undersize; the crank pin journals are offset-ground and reduced in diameter to 2 inch, to facilitate the use of the 1942 (21A) rods. A 3-3/8 inch bore and this crank (a 3/8 inch stroke over the stock Ford), results in the famous 3/8 × 3/8 296 cubic inch "killer Flathead motor".

Main Bearings

Remember to mark the main bearing caps before you disassemble the motor. If you should mix them up, here's a possible way of identifying them to avoid the need of align boring.

There are three main bearing caps. The rear one does not interchange with the two others. The bolt holes in the center cap are equally spaced between the sides of the cap, but the holes in the front cap are slightly offset. When installing the caps, align the bearing tang slots on the block.

Babbitt Bearings

If you have a '36 or earlier motor, the main bearings are the poured babbitt type. Babbitt metal is a soft, white, antifriction alloy

42

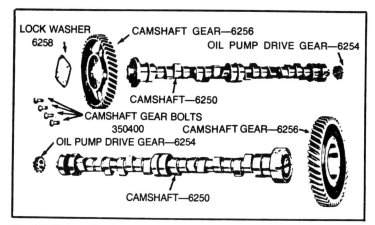

Fig. 4-7. Early and late interchangeable camshafts.

of tin, copper and antimony, named after Isaac Babbitt (1799-1862), the US metallurgist who discovered this alloy. Bearing babbitt is melted in a forge and poured into iron fixtures in the main webs and main bearing caps. The fixtures are jigged to the block and main caps to make a mold. When the babbitt cools, the jigs are removed. The machine shop will "mike" (measure with a micrometer) the crank you are going to use and then align bore the babbitt bearings to provide the proper clearance for the crank. Babbitt bearing motors can be cut to whatever clearance is desired.

Insert Bearings

Later crankshafts, '37-'53, use a conventional insert-type bearing. These bearings are available in the following sizes: .001 inch, .002 inch, .003 inch, .010 inch, .020 inch, and .030 inch. Replacement bearings are readily available from large manufacturers such as Federal-Mogul.

Camshaft

I stated previously that there were two types of camshafts used in the flathead. The earlier motors have a pressed-on gear and the later a bolt-on gear; both cams are interchangeable from motor to motor. (Fig. 4-7.) Camshaft journals should be 1.796 inch to 1.797 inch. Bearing clearance should be .001 inch to .003 inch. "Mike" the lobes of the camshaft and compare them with one another. As a rule, when a cam lobe loses its hardness, it wears fast

43

and it will be visibly smaller than the others. Replace a cam with differential measurements of more than .004 inch on the lobes or bearing journals. Replace a cam that is obviously damaged, worn, corroded, scored or that has discolored journals. Replace the drive gear if it is worn, broken or has chipped teeth. Replace the oil pump drive gear if it is worn, broken or has chipped teeth, or if it has been slipping on the camshaft.

If you find anything wrong, independent cam grinders can usually rebuild your cam at a better price than what the local machine shop can. Also, in the case of most of these motors, new cams are readily available from independent cam grinders and NOS from antique Ford dealers. In the long run, it's usually easier and cheaper to buy a new cam. Camshafts are made of cast nodular iron for all motors, with the exception of the '32 motor, which used a forged billet cam.

Summary

Most of you will probably send the block and components to the machine shop and have them do the necessary measurements and machining. The inspection and machining procedures that I presented were for reference, so you would know what goes on in the shop once your parts are there; and, more importantly, what to expect and demand from your machine shop. In a nutshell, here's what we should send to the machine shop first time around and what should be done .If you can send everything at one time, you may get a better price and you'll only have to make one trip:

☐ **Block.** Bore and hone; deck alignment; align bore; valve seat inserts; cam bearings, miscellaneous crack and thread repairs.

☐ **Heads.** Checked and/or milled for flatness. This coincides with the deck alignment; miscellaneous crack and thread repairs.

☐ **Pistons and Pins.** The pistons should always go with the block when it is to be bored, along with the piston pins-new or otherwise.

☐ **Crankshaft.** Checked for straightness; machine the main and con rod journals for new bearings: install new dowel pins; install a new drive gear.

☐ **Connecting Rods.** Checked for straightness; align bore and fit new bearings; fit new piston pins and bushings; have the rods shot peened for added strength and durability.

☐ **Valves.** Ground to match new or reground valve seats.

44

☐ **Camshaft.** If you plan on using your old cam rather than a new one, send it along also. Have it checked for straightness, lobe wear and bearing journal size.

☐ **Magnaflux.** Block; heads; crankshaft; connecting rods, caps and bolts; main bearing caps and bolts . . . even new ones.

Chapter 5
The Oiling System

In my mind, the oiling system of any internal combustion engine is the "key" system to survival and longevity. Oil's most important job is to prevent wear and scuffing of moving, internal engine parts. A good motor oil will do this by coating every moving part with a protective film that reduces friction and keeps engine tolerances within prescribed ranges.

Oil, aside from being a lubricant, performs four other functions that are no less important than lubrication:

□ **Cooling.** The cooling system of your flathead (radiator, water pumps, etc.) is responsible for about 60 percent of the engine cooling; oil handles the other 40 percent.

□ **Cleaning.** Oil will help keep contaminants in suspension until they are filtered out and/or drained away when you change oil.

□ **Sealing.** To the unaided eye, piston rings and cylinder walls may appear absolutely smooth, but they're not. A microscopic examination would reveal a large surface of irregularities in the form of hills and valleys. These valleys are easy escape routes for vapors and gases during the compression and power strokes of the piston. Motor oil coats these irregular surfaces with a film that is perhaps as thin as .001 inch, but that is enough of a seal to prevent loss of power and economic operation.

□ **Thick and Thin.** The oil in your motor has to be thin enough to let the starter crank the engine, yet thick enough to provide proper lubrication, and thin again to be capable of circulating rapidly and constantly through your motor.

That's an awful lot to ask of a liquid that still sells for less than a buck a quart.

Oil Pump and Press Pressure Relief Valve

During our engine tear-down, we found the oil pump mounted on the cylinder block inside the oil pan at the rear of the motor. The pressure relief valve on early motors will be found in the block. Later motors will have an oil pressure relief valve built into the pump itself (Fig. 5-1). On these motors (up to early 1949), the

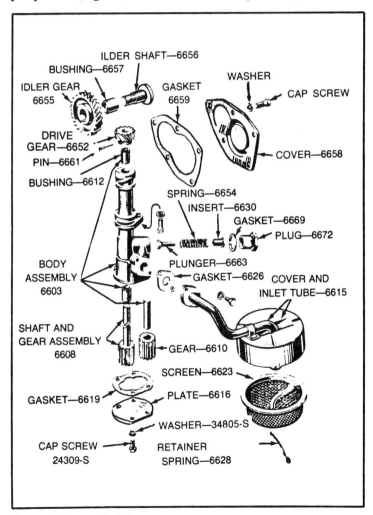

Fig. 5-1. Disassembled oil pump.

pumps had *spur* gears. From late '49 and on, an increased capacity oil pump was used, which had *helical* gears. There is no visible difference between the pumps. To determine which pump you have, remove the oil pump tube and screen and check the gears. While the spur gear pumps are adequate under normal conditions, I would recommend the use of a helical gear pump if at all possible.

Disassembly. Take that greasy old oil pump and soak it for several hours in a bucket of suitable solvent, to remove as much crud as possible. Once done, remove the strainer assembly retaining screws, the strainer and the gasket. Remove the cover plate and the pump driven gear. Remove the lock wire and the pressure relief valve (plug, gasket, spring and valve), if so equipped. Drive out the pin and remove the upper driven gear. Slide the shaft and drive gear assembly out of the housing. Once everything is apart, subject all the parts to another cleaning with solvent; wash and blow dry all oil passages in the pump body.

Inspection. Replace the oil relief valve spring if its tension is less than 78 ounces or more than 87 ounces when its length is compressed to 1.380 inch (1-9/16 inch). Replace the oil pump drive gear if it is worn or has broken teeth. Replace the drive gear shaft if it shows evidence of wear. Replace the oil pump body if it is cracked or broken. Replace the bushings if they measure more than .502 inch in diameter. To replace a bushing, simply drive the old one out with a suitable driver and press a new one in. Once installed, line ream the bushing to .500 inch. Measure the clearance between the pump gears and the pump body. It should be no greater than .005 inch. Replace the driven gear shaft if it measures less than .434 inch. To replace the shaft, drive it from the oil pump body and press a new shaft in place, making sure the lower end of the shaft will clear the oil pump cover when it is installed.

Replace the oil pump shaft and gear if it measures less than .434 inch. To reinstall the shaft in the oil pump body, position the shaft drive gear on the end of the shaft with the hub side of the gear down and the pin hole in the gear at a right angle to the hole (if any) already in the shaft. Press the gear onto the shaft until an end play of .017 inch is established. Drill a suitable hole in the replacement shaft to accommodate the holding pin (usually a 5/32 inch or ⅛ inch pin). Press the pin through the gear and the shaft.

Assembly. Apply a coat of 40 wt. oil to all moving parts and the internal passages of the housing. Install the oil pump driven gear on the shaft. Place the oil pump cover in position on the oil pump and install the lock washers and cap screws. Install the

48

screen cover and gasket on the pump body; install the two lock washers and cap screws that hold the screen cover to the pump body. Place the screen into the screen cover and install the screen spring retainer wire.

On the pumps having an oil relief valve, attach the relief valve spring to the plunger and insert them into the pump body. Install the relief valve nut or plug with gaskets as applicable; safety (lock) wire the nut/plug to the housing. The pump is now ready for installation on the motor ; a plastic bag will keep the pump "fresh" and clean while in storage, until the final engine assembly.

Oil Pan Removal and Replacement

For those not rebuilding your motor at this time and future reference for those who are, here are the procedures for removing

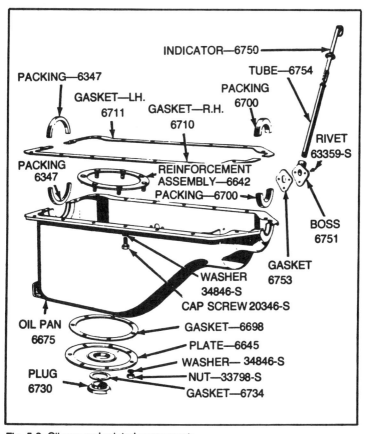

Fig. 5-2. Oil pan and related components.

and replacing the oiling system while the motor is still in the car (Fig. 5-2). While your motor may be running fine, a thorough inspection and cleaning of the system can do a lot for your peace of mind in the form of some good preventive maintenance, especially if your motor has been around a long time or logged a lot of miles since its last tear-down. It's also a good, worthwhile winter project, if you have a heated garage. While no two chassis are the same, the procedures are basically the same for all flathead motors.

Removal. Drain the crankcase. Remove the starter motor, clutch return spring and the flywheel housing front cover, if so equipped. Remove the bolt holding the road air breather duct (tube) and remove the tube. Remove the bolts holding the steering gear idler bracket to the frame. Remove the steering gear arm and drop the idler arm connecting rod until it hangs from the spindle arms. Remove the oil dip stick and unscrew the oil level indicator tube. Remove the oil pan retaining screws and the oil pan. On some cars, it will be necessary to disconnect the front motor mounts and lift the engine so that the oil pan will clear the front crossmember and crankshaft counter weights. Remove the oil pump assembly and proceed as outlined earlier in this section with the inspection and repair of the oil pump.

Cleaning. Wash the pan and related items in solvent, removing everything that is not oil pan: dirt, old gasket material and so on. Clean, as best you can, the bottom end of your motor.

Inspection. Check the oil pan for stripped threads, cracks, holes and warped gasket surfaces. Repair any irregularities as needed.

Oil Seal Replacement. Pry out the old packing in the front and rear seal retaining grooves. Install new packing in the recess at each end of the the oil pan. Soak the oil seals in light motor oil for two hours, then roll the packing in with a round bar or similar, making sure it seats properly in the recess.

Oil Pan Replacement. Make sure the gasket surface of the cylinder block is clean, and remove any burrs from around the threaded bolt holes. Tie each half of the gasket to the pan through two of the bolt holes to hold the gasket in place, while installing the pan. Hold the pan in place on the cylinder block and install two screws (not tight) in each side of the pan, but not the same holes as those used for tying the gasket. Remove the string ties and install the remaining screws; tighten them and torque them in place to 15-18 foot-pounds. Install the oil level indicator tube and dip stick, the road air breather tube and the flywheel housing front cover. If so equipped, align the flywheel housing front cover by installing the two shoulder bolts in the top holes.

Reinstall all of the items that were removed earlier; starter, clutch spring, steering idler arm support bracket, steering gear arm and tighten the motor mounts if they were loosened.

Odds and Ends

Your best bet is a new replacement oil pump. Your next best bet should be to rebuild your old pump, using new gears, shafts and bushings, along with a new oil breather cap. Here are a few odds and ends about the Ford oiling system that may or may not be of any value to you. Scoop type oil filter (breather) caps were used on the earlier models, atop the fuel pump. Later model Fords started using an oil filter as an accessory item, attached to the left cylinder head. Ford did not introduce an oil pressure gauge until 1935. Ford used 40 non-detergent oil in all of the flathead motors. The flathead oil system ran between 10-15 pounds of pressure, but is not considered a pressurized system by today's standards.

Chapter 6
The Cooling System

In the last chapter, I said the oiling system is the "key" system to survival and longevity in *any* internal combustion engine. The cooling system (Fig. 6-1) is no less important. In this chapter, we'll be looking at the flathead cooling system and rebuild the related components.

The Hot Head

The Ford Flathead is a notoriously hot running motor (for several reasons) and anything we can do to help keep these motors cool will be to our advantage (Fig. 6-2). In the overall scheme of things, there are some basics to watch for to avoid cooling problems:

☐ Check to see if one or both of the water pump impellers are installed backwards; the impeller blades go toward the pump housing and not the block. Also, check the bearings in the pumps. They should spin freely by hand with no drag.

☐ Be sure you have the correct thermostat for your car: bellows type for '40-'53; chemical type for '38-'39 and some '37's; and bi-metal type for '32-'36 and some '37's. The use of a 160-degree thermostat could also be of benefit.

☐ Check your ignition timing for a late (retarded) spark advance. Also try antifreeze in the radiator as a coolant.

☐ Make sure you have adequate oil pressure and oil of the correct viscosity.

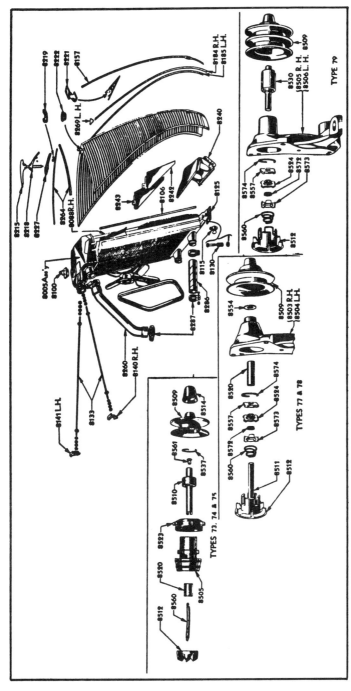

Fig. 6-1. Typical cooling system and related components.

53

☐ Water restrictors are of no use unless you're running a supercharger and/or a highly pressurized cooling system. If you have a pressurized system, try a new radiator cap. Don't laugh. To produce proper cooling in a pressurized system, the pressure must be contained and regulated. If the relief spring in the cap fails or you develop an air leak, the system will never reach its intended pressure level, the coolant will boil, and the motor will get hotter than a pistol.

☐ If you've got aluminum heads, beware of corrosion from electrolysis, which can block coolant flow. Anodizing the inside of the heads (water jackets) and the use of a water filter will help to control electrolysis.

☐ Is the fan belt loose, slipping, or oil soaked?

☐ Are your brakes adjusted too tight?

☐ Earlier in this project, I recommended that you have your block stripped by a commercial stripper, rather than sending it to the hot tanks. The few extra dollars you spent are well worth it. The block is immaculate and you can eat off those water jackets; there is no way that the hot tanks could get that rust and scale out of those jackets. Clean, unrestricted water jackets are essential to a cool running flathead motor.

☐ A clean radiator goes hand-in-hand with a clean block. Rust from the cast iron cylinder block, calcium chloride solution used for antifreeze or glycerine base antifreeze, will tend to form deposits in the radiator tubes, restricting the flow of coolant and the dissipation of heat. Have the radiator professionally rebuilt.

☐ When following the rebuild sequence (to follow), inspect your parts for cracks and wear; clean them thoroughly via the sandblaster or stripper and replace anything that is obviously in bad condition.

☐ The biggest problem most people have with their flathead cooling system, is the cylinder walls. Most flatheads will take a bore up to 0.125 inch (⅛ inch) and as these motors are bored out and meat is removed from the cylinder walls, heat is dissipated more rapidly to the coolant. As you bore your motor, you can expect (rule of thumb) temperature increases as follows:

stock to 0.030 over-175 F (normal)   0.090 over-195 F
          0.060 over-185 F   0.125 over-210 F

Cylinder sleeves to bring the bore back down will help; an electric fan, a variable pitch fan, and/or a fan shroud could be of benefit, if you've tried everything else.

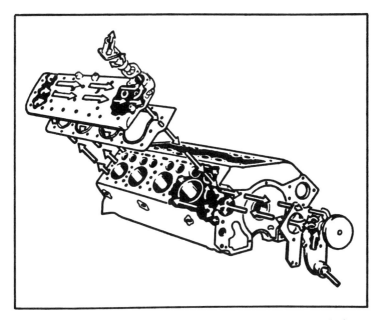

Fig. 6-2. Coolant flow. This is the '49-'53 block and is typical of all blocks.

□ What makes the flathead a hot head is that the exhaust gases must pass around the coolant in the water jackets, making these babies by nature run hotter than overheads. Is this bad and should you scrap the project? Not at all, on both counts, so long as you understand what you have and the characteristics of these motors. Heat is horsepower and as long as you don't boil over, you can safely operate these motors to 220 degrees F.

Water Pumps

There were three styles of head-mounted water pumps used on 1932-36 motors (Fig. 6-3). Two of the pumps are packless type having a porous shaft bushing which is lubricated by oil from a reservoir in the body (40 wt. motor oil.) Although these pumps are similar in design, there is a difference in them which should be noted. The original and early production pump used a coil thrust spring in the body which is made to bear on the impeller and shaft. In the later production and replacement pump, this coil spring is replaced by flat thrust spring and bakelite thrust washer. To disassemble, remove the pin that holds the pulley on the pump shaft and press the shaft and impeller out of the pulley; remove the

55

shaft and impeller from the pump body. The pump bushing should be replaced at this time. It, too, is a press fit.

To reassemble, first install the packing spring on the shaft with its smaller end next to the impeller, then the packing washer and packing, followed by the thrust washer. Insert the assembly into the pump bushing and install the front thrust washer. A new felt seal and retainer are then pressed into the pulley. The pulley and pump shaft are pressed together in such a way that the hole for the pulley retaining pin in lined up. Insert the pin and lubricator fitting and lubricate.

From 1937 to '48, the water pumps have been mounted on the front of the cylinder block instead of in the heads as in earlier models. The pump body is cast with a leg which is used as the front motor mount. There were two types of pumps used on these models. The impeller shaft of the first type rotates in a bushing, while in the second type, the bushing is replaced by a ball bearing which is part of the shaft assembly. Right and left pumps are not interchangeable as a unit. However, the parts within the pump housings are interchangeable.

These type pumps require no external lubrication. It should be noted that when removing and reinstalling these pumps, there is a mounting bolt located inside the pump inlet. To disassemble the bushing type, press the shaft out of the pulley and withdraw the impeller and shaft assembly from the rear of the pump. Remove the snap ring from the impeller and take out the thrust washer and seal assembly, noting the relative positions of the seal, clamp ring, spring guide and spring. The shaft is a press fit in the impeller. It should be replaced if its diameter measures less than .498 inch. If the bushing in the body of the pump is worn more than .502 inch on its inside diameter, press it out and install a new one; ream the bushing to .500 inch if necessary.

Mount the new seal assembly in the impeller in reverse order as noted when removing the old seal assembly. Insert the shaft, impeller and seal assembly into the bushing; install the front thrust washer and press the pulley on the shaft.

To disassemble the ball bearing type, support the pulley and press the pump shaft out of the pulley. Turn the pump over and press the shaft and bearing out of the pump body and impeller. Remove the snap ring, thrust washer and seal assembly from the impeller. Replace the bearing and/or shaft if it is worn or sticking. To reassemble, install the seal assembly and thrust washer in the impeller in reverse order as noted when removing the old seal assembly and lock in place with the snap ring. Press the shaft and bearing into the body of the pump, then press the pulley on the shaft until flush. Support the pulley and shaft and press the impeller on until it is flush with the end of the shaft.

Water pumps installed from 1949 to 53 are different in construction from previous years in that they are equipped with an oil cup for filling and lubricating with motor oil. To disassemble, remove the pulley from the shaft by using a puller. Remove the snap ring that holds the ball bearing assembly in the pump body. Use a puller to draw the impeller from the pump shaft and press the shaft and bearing out the front of the pump body. This bearing is a press fit on the impeller shaft. Press the bushing and seal assembly out of the pump body and install a new bushing and seal.

To reassemble, press the bearing on the shaft, until it touches the retaining snap ring. Insert the assembly through the front of the pump body and press the shaft through the bushing and seal: Use the outer race of the bearing to press the shaft into place. Install the second bearing retaining snap ring and press the pulley onto the shaft, until it seats against the inner race of the bearing. Press the impeller on the shaft, allowing a clearance of .030 inch-.040 inch between the impeller blades and the pump body. Fill the oil cup

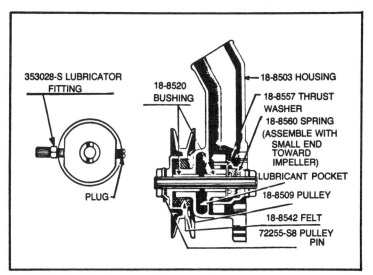

Fig. 6-3. Early style water pump. These are adaptable to the right or left bank by changing the lubricator fitting from one side of the pump to the other. Use ordinary chassis lube when lubricating this pump, as high melting point greases are insoluble in hot water and may clog the radiator if the pump is over-lubricated.

with 40 wt. motor oil, making sure the felt wick is completely saturated.

Thermostats

Three types and four styles of thermostats have been used since 1932 (Fig. 6-4). In all models through 1936, the bi-metal, hose-mounted type is used. In 1937, the bi-metal type was re-designed for head outlet mounting; also in 1937, some production models used a chemical type. From 1938 through 1939, the chemical type was used. From 1940 through 1953, the bellows type was used.

Several things can cause improper thermostat action:

☐ In early models, the clamp around the thermostat used to hold it in position in the hose could be too tight, causing the valve to stick.

☐ Hose mounted thermostat was installed upside down.

☐ Head outlet mounted thermostat was not held securely in place. The shoulder on the inside of the hose provides a means of holding the thermostat in place. If the hose is not forced down tightly on the head outlet, there will be sufficient room to permit the thermostat to rock, allowing water to by-pass the thermostat. In addition, this rocking action strikes the coil supports against the sides of the outlet housing, which, in time, could affect the characteristics of the thermostat.

☐ Dirt and pieces of rubber hose imbedded on the edge of the butterfly, holding the valve either open or closed.

The following procedure can be used to check the action of the thermostat: Heat a pan of water to 190 degrees F, then place the thermostat in the water. It should open immediately. To check its operating range, allow the water to cool slowly, stirring constantly to maintain even temperature. When the water gets to 180 degrees, the thermostat should start to close and should be fully closed when the water temperature has dropped to 145 degrees. A recheck can be made by reheating the water after the temperature has dropped below 145 degrees and observing the action of the thermostat. It should be noted that replacement units sometimes have different opening and closing points. The temperature range is usually stamped on the thermostat body. An example of this would be the '49-'53 type, which starts to open at 160 degrees and is fully open at 175/180 degrees.

Summary

If all else fails with the rebuilding of your water pumps, rebuilt and NOS pumps are readily available. A little tip I might pass along

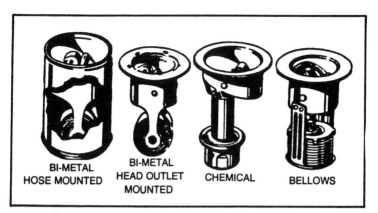

BI-METAL
HOSE MOUNTED

BI-METAL
HEAD OUTLET
MOUNTED

CHEMICAL

BELLOWS

Fig. 6-4. Thermostats used in Ford motors from 1932 to 1953.

when mounting your pumps to the block ('37-'53) is to use a stud and a nut, rather than a bolt, on the water inlet tie-down. It's a dumb place to have put a bolt, but it's there and we have to live with it.

The stud and bolt combination will make it easier to remove the pump somewhere down the road, when it's rusted together again. A little dab of silicon RTV around the nut when installed and torqued into place will form a protective shield and also make life easier, if and when that pump ever has to come off again.

Chapter 7
The Fuel System

All Fords from 1932 are equipped with a rear-mounted fuel pump operated by the camshaft. Since 1935, all V8 engines use a dual barrel carburetor; V8's prior to 1935 use single barrel carburetors (Fig. 7-1).

Manifolds

The dual intake manifold used on V8 engines in conjunction with the dual downdraft carburetor distributes the fuel-air mixture to cylinders in the order shown in Fig. 7-2. To provide proper fuel vaporization, small passages in the bottom of the manifold casting line up with passages to the exhaust system in the cylinder block, allowing the gases to flow through the heating chamber of the manifold, warming it and the fuel within its passages. Original manifolds were made of aluminum but the cast iron type was introduced later to prevent heat dissipation.

Fuel Pump

The fuel pump on the V8 engine is mounted at the rear of the intake manifold and operated by a push rod which, in turn, is operated by the camshaft.

When it is determined that the carburetor is not receiving fuel, check the following: fuel level in tank; filter screen in fuel pump; remove the cover from the pump to inspect the screen; cork

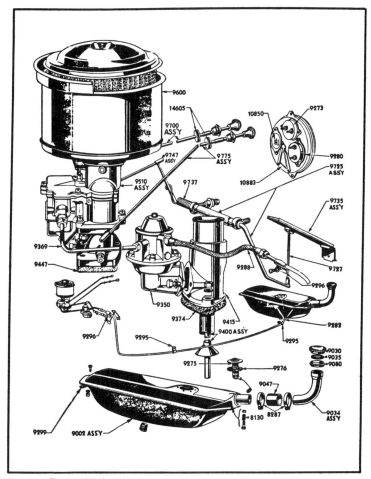

Fig. 7-1. Typical V8 fuel system.

gasket in pump cover may be improperly seated; fuel line leak; numerous cases of pump failure are traced to loose connections at the fuel tank, the pump or a deteriorated flexible fuel line in which case the pump will draw air instead of fuel; clogged fuel line.

If the above inspection procedure does not locate the trouble, it is possible to perform a few checks on the pump itself. Take off the pump cover and screen. Inspect inlet and outlet valves. Replace valve gaskets if necessary and tighten valve plugs or screws. In case of a leak at the diaphragm, tighten the ring of upper body screws.

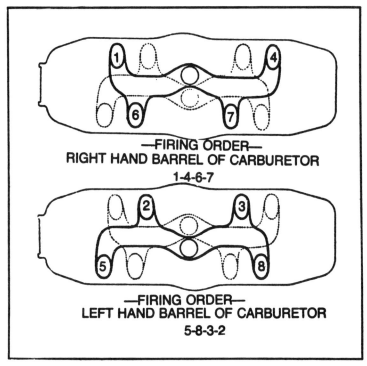

—FIRING ORDER—
RIGHT HAND BARREL OF CARBURETOR
1-4-6-7

—FIRING ORDER—
LEFT HAND BARREL OF CARBURETOR
5-8-3-2

Fig. 7-2. Intake manifold distribution of the fuel and air mixture.

Any other repairs to the fuel pump must be done with the pump removed.

Disassemble the pump and discard the items that will be replaced by those in the repair kit. Clean up the remaining parts and assemble the pump as follows: Insert the link, rocker arm and rocker arm bushing in position and secure them with the rocker arm pin. Place diaphragm spring and diaphragm into lower body and hook the diaphragm pull rod on the link. Install the rocker arm spring in lower body. Lock the valves, plate and gasket into upper body with the two valve screws. With rocker arm held in the up position, place the two halves of pump together and install the six body screws loosely. Release the rocker arm and tighten the screws evenly. Next, install screen, gasket and cap and secure with the washer and cap screw.

Suction test for a fuel pump can be accomplished in the following manner: Attach a hose or line from fuel pump inlet to a fuel container and operate the rocker arm manually. A good pump should raise the fuel at least 30 inch with a maximum of 40 strokes.

With a vacuum gauge available, the pump should show a reading of 10 inches of vacuum.

Fuel pump pressure to carburetor when tested with a fuel pressure gauge should not be less than 1 ½ pounds nor more than 3 ½ pounds. Fuel pump pressure on V8s can be decreased by adding thick gaskets between intake manifold and the combination pump adapter, oil filter, and crankcase breather. Pump pressure can be increased by placing a spacer between the push rod and pump rocker arm socket.

Carburetors

Early V8 engines are equipped with a single barrel downdraft carburetor. The fuel level in this carburetor should be 1-3/16 inch, plus or minus 1/16 inch measured from the top of the fuel to the top of the float chamber. Engine idle speed is controlled by the throttle plate adjusting screw (Fig. 7-3). Idle fuel mixture is controlled by the metering pin on the top of the carburetor. Turning the pin clockwise leans the mixture, and counter-clockwise richens the mixture. Proper procedure for adjustment is to adjust idling speed to the equivalent of 5 mph, then turn metering pin clockwise until the mixture is so lean that the engine operates roughly. Open metering pin about ½ turn at a time until engine runs smoothly. The initial cold setting is to turn the metering pin in until air vanes are just starting to open, then turn pin out 5 turns.

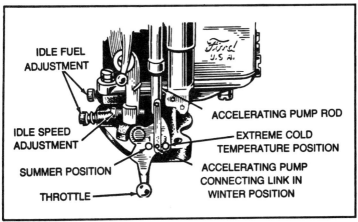

Fig. 7-3. Carburetor adjustments. Adjust the idle speed to 475-500 rpm. Adjust the idle mixture screw until the engine starts to miss, then back the screw out until it runs smoothly. Note that these carburetors have two idle mixture screws, both of which must be adjusted to balance the engine at idle speed. Manifold vacuum should be 18-21 inches (of mercury) with the engine idling.

Dual carburetors for V8 engines (Fig. 7-4) can be adjusted for idle speed and idle mixture only. Fuel level for Stromberg carburetors (marked on the side with "97" for 85 hp engines and "81" for 60 hp engines) should be 15/32 inch plus or minus 1/32 inch measured from top of fuel to top of float bowl or underside of cover gasket. Fuel level for Ford carburetors should be 11/16 inch below the top surface of bowl. Change fuel level by bending float arms, evenly, so that the float remains level.

To check and adjust Ford carburetor float level with the air horn (bowl cover) off, invert horn with float attached, and float valve in its seat. Measure from the bottom of the float (not the soldered seam) to the gasket surface of the air horn. This should be 1 ⅜ inch to 1 11/32 inch and can be obtained by bending the float arms as explained above. This operation will give correct fuel level for Ford carburetors.

Idling speed is set by the throttle stop screw. Mixture is adjusted by two needle valves in the carburetor body (one valve on single carburetors). Turning valves in leans the mixture, out richens the mixture. Proper procedure is to adjust idling speed then turn one needle in until the engine slows down, then out slowly until the engine runs smoothly. Adjust second needle in the same manner, then readjust the first needle to obtain smoothest action. If the above procedure does not correct rough idle, or failure to idle, (and there is no doubt that the trouble is not due to ignition or leaky valves or vacuum leak) remove the idle needles and clean them. If this doesn't remedy the situation, dismantle the carburetor and clean thoroughly.

Set the accelerating pump link in the proper hole depending upon climate and season. On the type with three holes, one is marked S for summer; one is marked W for winter and the center hole is for intermediate operation. On the types with two holes, the shortest stroke is for summer operation and longest stroke is for winter operation.

Fuel Gauges

The hydrostatic fuel gauge (Fig. 7-5) is used on Ford cars up to 1935 inclusive and it operates in the following manner: Fuel in the tank causes air in the sending unit and in the connecting line to the dash panel unit to be compressed, and in turn forces a red liquid up in the tube of the dash panel unit. When the system is in working order, the height of this liquid indicates the relative amount of fuel in the tank.

An inoperative gauge can be caused by a clogged or leaking air line, lack of fluid in the dash panel unit or a faulty tank unit. To check the dash unit, remove it from the car and apply a little air pressure to the unit by moving your thumb up and down rapidly against the air line connection on the back of the gauge. Hold the pressure obtained with the thumb over the opening and note the height of the liquid in the tube. If liquid holds at the same level until thumb is released, the unit is in working order. If the liquid will not rise in the tube, there is either an air leak in the unit or the tube is plugged.

With no air pressure on the unit, the liquid level should be at the bottom (zero) line. If fluid level is low, add additional amount with a medicine dropper through the air line connection. If fluid level is too high, remove the excess by inserting a toothpick or pipe cleaner into the air line connection.

If an attempt is made to blow out the air line, disconnect both ends and use filtered dry air as ordinary compressed air contains moisture which will affect gauge reading.

Whenever the gauge, line or sending unit is disconnected, the system balance is disrupted and will require that the following to be performed when reconnected: Disconnect fuel line at the fuel pump and blow into it with the mouth. The air which is forced into the fuel tank will then enter the sending unit, displacing the fuel which entered when the connection was broken.

This balance would return eventually with the natural motion of fuel in the tank, but it may take a week or more depending on the amount of driving done.

The electric gauge used on all models after 1935 consists of a tank sending unit and a dash indicating unit connected by an insulated wire. The dash unit, in turn, is connected to the ignition circuit. Never allow the gauge reading to progress past the ¾ mark if the wiring between the dash unit and the tank unit should become grounded.

Modifications for 1949

The manifolding on the 1949 engines is similar to previous models. The exhaust cross-over pipe on the V8 has been rerouted around the front of the engine and under the radiator hoses to permit easier removal of the oil pan. The V8 intake manifold has been redesigned so that the crankcase breather and the oil filler pipe is located between the generator mount at the front, and the

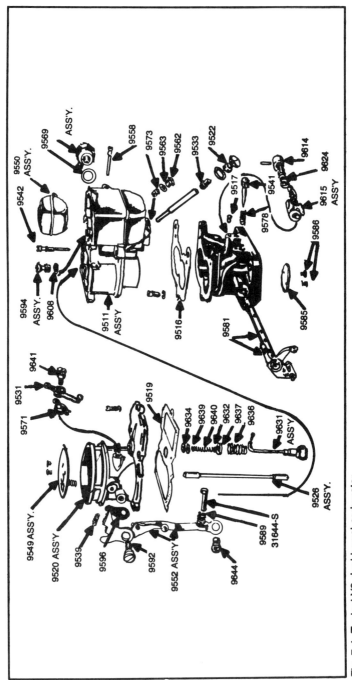

Fig. 7-4. Typical V8 dual barrel carburetor.

66

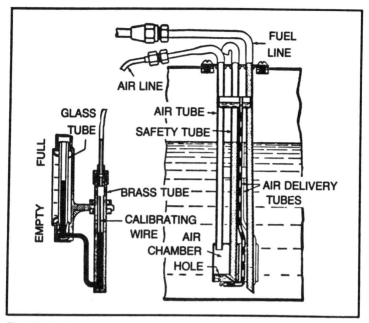

Fig. 7-5. Hydrostatic fuel gauge and sending unit.

carburetor. The fuel pump mount has been changed to a different angle. These new positions are for better accessibility.

Fuel is delivered to the carburetor under higher pressure than in previous models. The fuel pump on the SIX produces a pressure of 4 to 5 pounds, while the pump on the V8 produces 3 ½ to 4 ½ pounds. Because of the increased pressure, and also to prevent vapor lock, the fuel lines have been increased to 5/16 inch.

An electrically operated fuel gauge is employed. The sending unit in the tank is accessible through an opening in the floor of the trunk. The carburetors are designed for use with the vacuum distributor used on these models. As will be explained in the ignition section, the carburetor has passages which open into the venturi and throat of the carburetor just above the throttle plate.

Note: Do not attempt to use earlier carburetors with 1949 ignition system.

Carburetors are adjusted in the same manner as described for earlier models. Three holes are provided for positioning the accelerating pump link. The inner hole nearest the shaft is the summer setting, center hole is the intermediate setting, and the outer hole is for winter operation.

Chapter 8
The Distributors

The '32-'48 distributors are mounted on the engine front cover and driven directly by the camshaft (Fig. 8-1). The coil is mounted on top of the distributor body except on later models, in which the coil is mounted separately. This system operates on four volts with a resistance unit (located on the fuse or circuit breaker block under dash) is inserted in the low tension lead from the ignition switch to obtain the correct voltage. Do not operate the Ford ignition (Figs. 8-2 and 8-3) on six volts .

There are three adjustments provided on these distributors: breaker point gap, initial spark advance and vacuum brake control. The setting of the breaker point gap can best be performed with unit removed from the engine. The vacuum brake is adjusted after initial spark advance setting is made, and with the distributor mounted on the engine (Table 8-1).

The Early Models

To remove the distributor, disconnect the low tension lead, the vacuum line, and unfasten the distributor cap (or caps if it is an earlier model). The distributor-to-front-cover cap screws can now be removed and the distributor taken off. With the earlier models, it may be necessary to remove the fan and fan belt to remove the distributor.

Before installing the distributor recheck the breaker point gap. The following procedure will speed up the installation of the

Table 8-1. Distributor Tune Up Data.

Year	Engine	Spark plug gap	Breaker point gap	Cam angle in degrees	Firing order	Compression pressure at cranking speed	Breaker spring tension
1932-36	V8/85	.025"	.013"	35	15486372	105 psi (95 psi on '32 motors)	22-27 oz.
1937-38	V8/85	.025"	.015"	36	15486372	100 psi	20-24 oz.
1939-42	V8/85	.025"	.015"	36	15486372	100 psi	20-24 oz.
1946-48	V8/100	.025"	.015"	36	15846372	100 psi	20-24 pz.
1949-53	V8/100	.030"	.015"	29	15846372	92 psi	not applicable

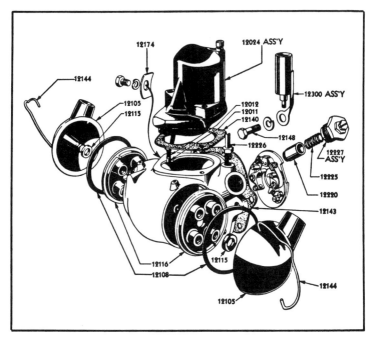

Fig. 8-1. Early style V8 distributor.

distributor. Feel the groove in the camshaft to determine its position, then turn the distributor shaft tongue to the same relative position, remembering that the tongue on the shaft and the groove in the camshaft are off-center so the distributor can be installed in only one position.

Install the distributor gasket on the distributor housing using a quick-drying gasket cement (very early models also had a second gasket which is installed in the groove on the front cover). Place the distributor on the front cover making no attempt to position the tongue of the distributor shaft in the camshaft groove. Install the distributor-to-front cover cap screws just enough to grip their threads. Rotate the distributor shaft (by turning the rotor) forward and back until the tongue aligns with the camshaft groove, then push distributor into place and tighten the can screws.

When correct distributor alignment is obtained, it will slip into place easily, so do not attempt to draw the distributor into position with the cap screws. After installation, adjust the vacuum brake as described in the following paragraphs.

Distributors manufactured for 1937-40 engines contained four mounting holes, two of which are close together. This distributor

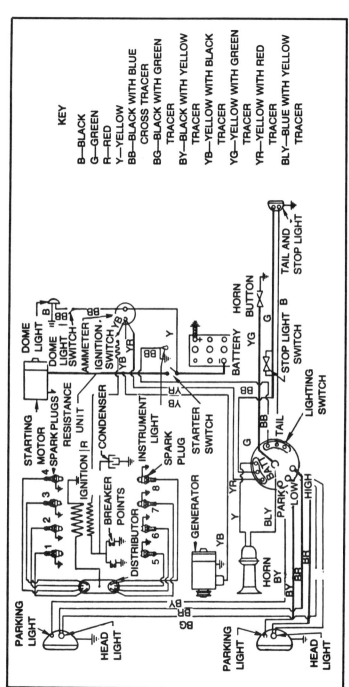

Fig. 8-2. Typical 1932-1935 V8 ignition schematic.

71

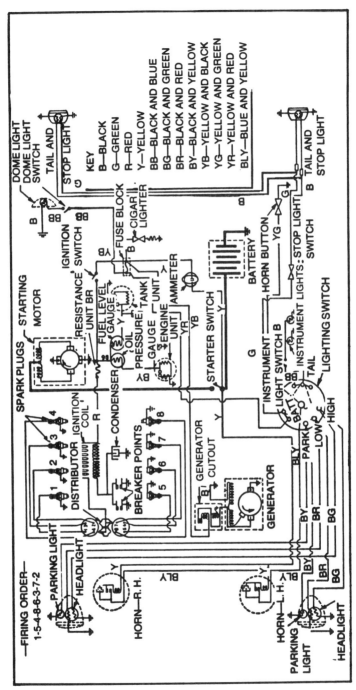

Fig. 8-3. Typical 1936-1948 V8 ignition schematic.

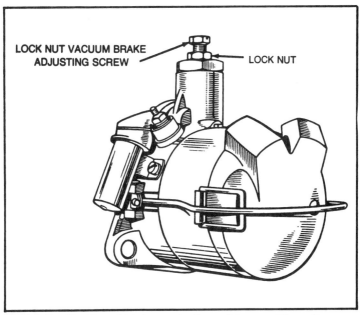

Fig. 8-4. Road test the vehicle by accelerating rapidly several times. If you can hear a "ping" or detonation, tighten the vacuum brake on the distributor until the ping disappears.

is used on both 85 and 60 hp engines. When installed on the V8-85 engine, one of the two adjacent holes is used and when installed on the V8-60 engine, the other mounting hole is used. There is no difference in adjustment or parts on this distributor when used on either engine.

Adjust the breaker point gap by means of the stationary point to the value given in the Tune Up Table. On the V8 distributor, adjust both sets of points to the same value. Be sure the breaker arm rubbing block is on high point of cam when making the adjustment. In order to locate any worn cams check the gap with breaker arm rubbing block on each cam; make the adjustment on earlier models with the coil mounted in place.

The usual method of setting initial spark advance is with a special timing fixture. However, two other methods which require no special equipment can be used.

Under the first method set the distributor advance plate with the unit off the engine. With the breaker point gap correctly adjusted, place a steel rule, or straight-edge, on the wide side of the tongue on the distributor shaft. Place another rule, or a small square, working from the mounting hole nearest the timing adjustment screw. Rotate the distributor shaft until the edges of the two rules touch. The left-hand breaker points on the V8 model should just be starting to open. Adjustment to obtain this condition is by movement of the advance screw and plate on right side of distributor. Move up to advance and down to retard. Always repeat timing procedure after an adjustment. Always turn the distributor shaft backwards at least 1¼ turn, then forward to eliminate backlash, before repeating the timing check.

The second method of setting initial spark advance without special tools is to do so with the distributor mounted on engine. Remove the spark plugs from the first three cylinders of the right bank; turn engine until number one piston is rising on the compression stroke (this can be determined by holding your thumb over spark plug hole). Using a graduated measuring rod or stick, continue to turn the engine until numbers two and three pistons are precisely the same distance from the top of the cylinder. When two and three pistons are thus positioned, number one piston is on top dead center. Now, turn on the ignition switch and retard the spark by pushing down the advance plate on the right side of the distributor. Hold number one plug wire close to the cylinder head and slowly move the advance plate up until spark occurs at the gap between the wire and head. Push the advance plate up one more graduation to obtain the necessary 4 degree spark advance.

Any difference in timing made necessary by the use of different grades of fuel or changes in compression ratio can be established by adjusting the vacuum brake (Fig. 8-4). With the distributor installed, back off on the vacuum brake screw until the engine "pings" under load, on a road test. Then turn the screw in until the "ping" is removed and tighten the lock-nut to maintain the adjustment. Sometimes an inoperative vacuum brake is caused by the lack of a washer on the end of the adjusting screw. This allows the screw to enter the center of the vacuum piston spring instead of applying tension to it; the vacuum brake should not be installed without this washer.

Later Model Distributor

On the '49-'53 distributors (Figs. 8-5 and 8-6) a vacuum diaphragm connected to the carburetor operates the breaker plate for automatic spark advance. There are no centrifugal weights or vacuum brakes used on these units.

Loosen the lock screws holding the stationary point, and place a screwdriver of proper width blade in the adjustment slots. Move the point to correct spacing (breaker arm rubbing block must be on high point of cam). Point gap should be .014 to .016 inch. Retime the ignition after the breaker point gap adjustment. Spark plug gap is .030 inch.

The distributor can be timed on the engine and the procedure is quite simple: The mark on the crankshaft front pulley must be aligned with the timing pointer. Remove the number one spark plug and turn the engine until the piston is rising on the compression stroke (a rush of air will be felt as the thumb is held over the plug hole). Continue turning until the timing mark on the pulley aligns with the pointer as previously described. Turn the ignition switch on, loosen the distributor lock plate screw, and turn the unit counter-clockwise about 1⅛ turn. Hold the number one plug wire close to the cylinder head and turn the distributor clockwise slowly until a spark occurs at end of plug wire. Lock the distributor in this position.

The advance system in this distributor is controlled by calibrated springs and vacuum; there is no centrifugal advance mechanism. The eight-lobe cam operates a single set of breaker

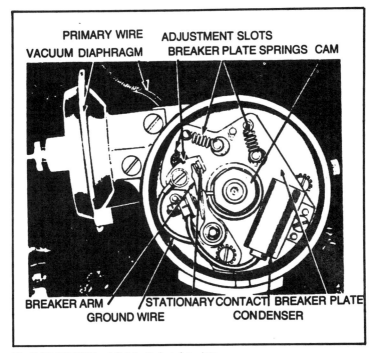

Fig. 8-5A. '49-'53 Ford distributor breaker plate.

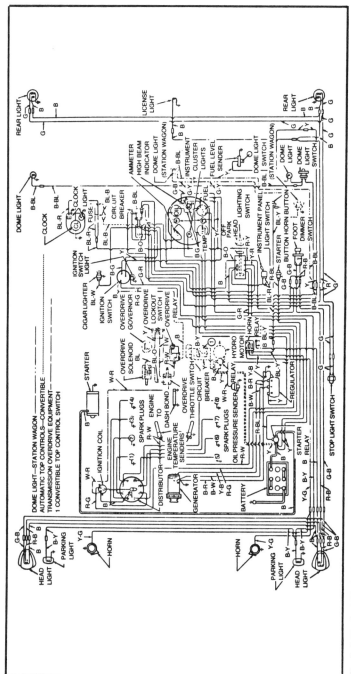

Fig. 8-6. Typical 1949-1953 V8 ignition and electrical schematic.

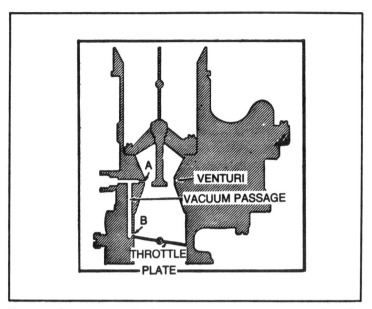

Fig. 8-7. Cross-section of '49 carburetor showing vacuum openings for distributor operation.

points. Passages in the carburetor (Fig. 8-7), open to the venturi and carburetor throat just above the throttle plate. The average or mean vacuum at these two openings acts on the diaphragm of the distributor. This vacuum varies with throttle opening and road load. During acceleration, the vacuum in the venturi opening increases but the vacuum at the throttle plate opening (which is actually manifold vacuum) decreases. The passage between the two openings balances the variation. Thus the vacuum at the distributor lowers, and the calibrated breaker plate springs retard the advance from its road load setting. As speed is increased (and load decreases) the vacuum at the two openings increases and overcomes tension of the springs, bringing the breaker plate back to advance position. During part throttle operation, the vacuum is high and the spark will be fully advanced. The two calibrated springs are precision set at the factory and require special stroboscopic distributor equipment to adjust them. Most tune up shops can handle this adjustment on their Sun machine; cost is minimal.

Note: Never connect the distributor vacuum line to intake manifold vacuum.

77

Chapter 9
The Electrical System

The electrical system of a motor consists of a battery, starting motor, control switch, generator and a voltage regulator (Fig. 9-1.)

Starters

The starting motor incorporates a Bendix drive which eliminates the necessity for a pedal or solenoid-operated engagement of the starter pinion with the flywheel ring gear (Fig. 9-2 and Table 9-1). On models prior to 1937, the starter switch is mounted on the floor board and the heavy cables from the battery and starter are connected directly to it. When the battery position was changed from under the floor boards to the engine compartment, a solenoid switch was added to eliminate the lengthy cables and provide a more direct contact between the battery and starter.

An inoperative starting motor may be caused by one or all of the following:

☐ **Starting motor operates but does not turn engine:** A broken Bendix spring or gummed pinion shaft, preventing the rotation of the pinion toward the flywheel ring gear.

☐ **Motor does not operate when switch is closed:** Improper contact of switch, solenoid, or starter button. Corroded or loose battery cable or starting motor cable. Short circuit in starting motor.

☐ **Poor brush contact or bad commutator surface:** Seized engine.

☐ **Pinion engages flywheel ring gear but fails to disengage:** A bent starting motor shaft or damaged flywheel ring gear teeth.

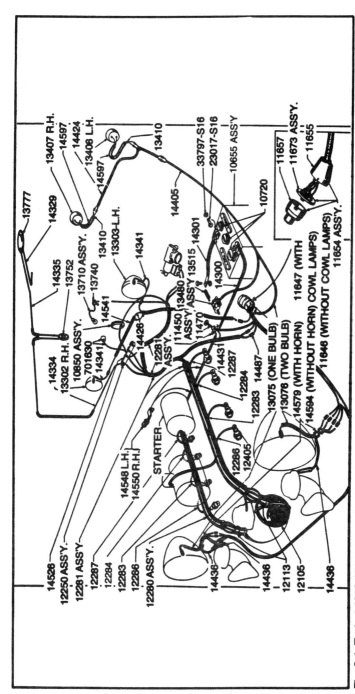

Fig. 9-1. Typical 1932-1934 electrical schematic.

79

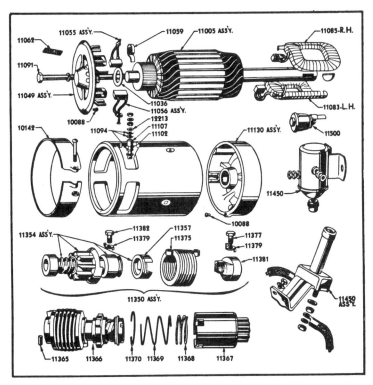

Fig. 9-2. Basic Ford starter.

Generator Systems

Early model Fords are equipped with a low output generator and cutout. With the installation of radios and other electrical accessories it was necessary to use a higher output generator with a step voltage control type regulator to control the maximum output. A still higher output generator with regulation for both voltage and current was used in '49-'53 models (Figs. 9-4 and Table 9-2). Whenever replacing a generator or its control device, use the proper replacement item. For example: a cutout cannot be used alone with a two-brush generator as the generator would burn itself out having no means of controlling the field current. The Generator Control Specifications Table gives generator control specifications.

Third Brush Generator with Cutout. The simplest system in use, and installed on 1932 to 1934, cars is the low output generator having an adjustable third brush which maintains correct output, and a cutout to prevent discharge of the battery through the

Table 9-1. Starter—Basic Ford Part Numbers and Nomenclature.

10088—Dowel	11354—Shaft and Pinion Assembly
10142—Band	11357—Drive Shaft Sleeve
11005—Armature	11365—Pinion Drive Pin
11036—Armature Thrust Washer	11367—Starter Drive Pinion and
11049—Brush End Plate	Head assembly
11055—Field Brush	11368—Meshing Spring
11056—Armature Brush	11369—Anti-Drift Spring
11059—Brush Spring	11370—Pinion and Head Retainer
11083—Left Hand Field Coil	Ring
11085—Right Hand Field Coil	11375—Starter Drive Spring
11091—Through Bolt	11377—Screw
11094—Washer	11379—Lock Washer
11102—Field Contact	11381—Starter Drive Head
11107—Field Terminal Bushing	11382—Screw
11130—End Plate	11450—Starter Switch Assembly
11350—Starter Drive Assembly	11500—Starter Switch Push Button
	12213—Washer

generator when the engine is not running, or when the generator charging rate is lower than the battery charge. To adjust the third brush, move it in the direction of armature rotation to increase the charging rate, and in reverse direction to decrease the charging rate.

Third Brush Generator with Cutout and Step Voltage Control. The generator used in this system is of higher output (about 18 amperes) than earlier models, and has a step voltage control to reduce the charging rate to about 20% when the voltage reaches 8.3 volts. A cutout relay is included with this control and has the same function as the above unit, which is to prevent the battery from discharging through the generator when the generator output is lower than the battery charge. Again, the generator output is adjusted by means of the third brush.

Table 9-2. Generator—Basic Ford Part Numbers and Nomenclature.

10005—Armature	10121—Washer
10051—Brush Holder	10122—Oil Seal
10057—Brush Spring	10129—Brush End Plate
10062—Brush Holder Insulator	10130—Pulley
10069—Main Brushes	10134—Washer
10070—Thrid Brush (generator field	10135—Bearing Adjusting Collar
control)	10136—Bearing Lock Ring
10072—Third Brush Plate Assembly	10139—End Plate and Bracket
10088—Dowel	10141—Oil Cup
10094—Bearing	10142—Band
10100—Lead Wire	10175—Generator Field Coil
10104—Generator Lead Wire Grom-	10193—Field Coil Connecting Wire
met	Insulator
10112—Bearing Packing	10505—Cutout Assembly
10113—Felt Bearing Retainer	
10120—Through Bolt	

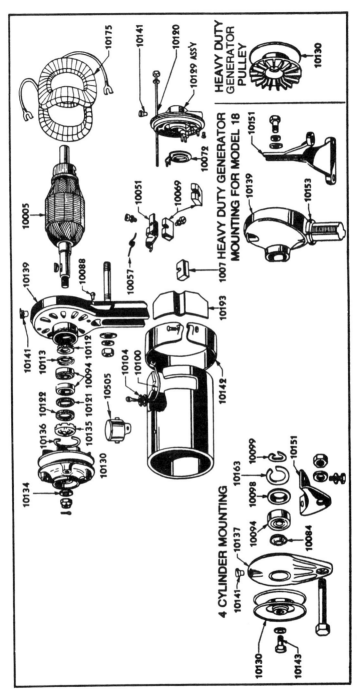

Fig. 9-3. Basic Ford generator.

Table 9-3. Generator Control Specifications

Part number	Type	Cut in voltage		Voltage regulation at 70 F.		Amperage regulation at 70 F.	
		Min.	Max.	Min.	Max.	Min.	Max.
B-10505	Cutout only	6.1	6.3	-	-	-	-
01A-10505	Standard	6.1	6.3	7.0	7.2	30	33
*68-10505	Two rate relay	6.1	6.3	8.0	8.3	*	*
59A-10505	Standard	6.1	6.3	7.0	7.3	30	33
5EH-10505	Standard	6.1	6.3	7.0	7.3	39	42

*Reduces charging rate when voltage becomes excessive (3 brush generator only).

Two Brush Generator with Voltage and Current Regulator. This system consists of a two brush generator having its field circuit and output controlled by regulators. The regulator assembly contains three main units; the cutout; the field current regulator; and the voltage regulator. This system can produce a maximum of about 30 amperes and all adjustments are made within the regulator assembly.

Polarizing the Generator. When installing a generator, it is *extremely* important that the generator be polarized, so that current flow will be in the proper direction.

To polarize your generator, momentarily connect a jumper lead between the "Gen" and "Batt" terminals of the regulator. This allows a momentary surge of current to flow through the generator, which correctly polarizes it. Failure to do this may result in severe damage to your electrical equipment since reversed polarity causes vibration, arcing and burning of the relay contact points within the regulator.

Chapter 10
Engine Reassembly

The block should be given the hot or cold tank cleaning treatment again. After this, the newly machined surfaces should be thoroughly washed with ordinary soap and warm water. All oil passages in the block should be cleaned out with a rifle brush and blown clear with compressed air. This is a vital operation, since any small particles of metal or dirt trapped in an oil gallery can cause engine failure. After cleaning, all of the machined surfaces should be coated with a light oil to inhibit flash rusting.

The engine should be painted with heat-resistant paint. The inside surfaces that are not machined may be painted with a light coat of Rust-Oleum brand paint. This will tend to keep sludge from finding a foothold on the rough casting (this is a very common practice in race built motors). A green plastic trash bag will keep the motor clean when not being worked on. Before you begin the reassembly sequence, make sure that all of the component parts are in satisfactory condition for use, as outlined in Chapter 4—Inspection and Machining (Fig. 10-1).

Making Sure The Parts Fit

Plastigage has been around a lot longer than I have and is a very handy, inexpensive tool when it comes time for engine assembly work. When used correctly it can provide amazingly accurate results. For our application, we will use plastigage to check out main and rod bearing clearances. Plastigage consists of a wax-like plastic material which will compress evenly between the bearing and journal surfaces, without damaging either surface. Plastigage is color-coded and we will use "green". Green is used for checking clearances between .001 inch and .003 inch, which is where our tolerances fall.

Work with the motor upside down, and work only on one bearing at a time. Clean the bearing and journal and lay a piece of plastigage across the width of the journal. Install the bearing and cap and torque them to the required specs. Loosen the bolts and remove the cap. The plastigage will be smashed against either the bearing insert or the crank journal. Using the scale on the wrapper, compare it to the width of the crushed plastigage. The closest matchup will indicate the amount of clearance. The scale is calibrated .003 inch, .002 inch, .0015 inch, .001 inch.

The Crankshaft

Select main bearings of the correct thickness to establish a clearance of 0.001-inch to 0.003-inch. If the wear on the crankshaft is such that the clearance cannot be obtained with the bearings available, it will be necessary to remachine the crankshaft to the next undersize for which bearings are available.

If a cylinder block (part number prefix 41A) having a main bearing bore of 2.670 inches to 2.671 inches is to be used with a 1937 or 1938 crankshaft, use special bearings having a larger outside diameter of the correct thickness to obtain the correct bearing clearance. In addition to the above combination, a small percentage of cylinder blocks were manufactured with a 0.015-inch oversize main bearing bore. These blocks can be identified by the letters "ERP" stamped on the gasket surface for the oil pan at the front left-hand side of the engine. These blocks, when new, had a metal tag attached, indicating the block has an oversize main bearing bore. Bearings have an 0.015-inch oversize outside diameter are required for these blocks.

Lubricate and install the three upper halves of the main bearings in the cylinder block. Lubricate and install the lower halves of the main bearings in the main bearing caps. If the old main bearings are being reused, assemble them in the cylinder block and bearing caps in their original position as indicated by the markings made during the disassembly procedure.

Install the upper half of the retainer in the cylinder block. Engines built since April 1, 1944 have the lower oil seal retainer

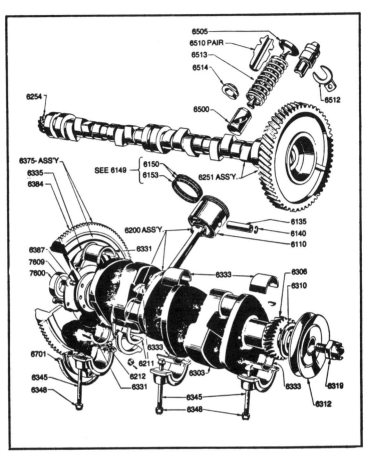

Fig. 10-1. Typical V8 reciprocating assembly.

integral with the cap. On earlier engines, install the lower half of the retainer in the rear main bearing cap. Oil the main bearing inserts with a light coat of oil. Install the oil slinger on the forward end of the crankshaft. Place the crankshaft in the cylinder block and install the main bearing caps on the cylinder block. Install the main bearing cap nuts and tighten them to from 75 to 80 foot-pounds. Pry the crankshaft forward and insert a feeler gauge between the crankshaft and rear main bearing (Fig. 10-2). If the clearance exceeds 0.008 inch, select a bearing with a thicker flange or if the clearance is less than 0.002 inch, select a bearing with a thinner flange. Lock all the main bearing nuts with wire.

In another method used to check main bearing clearance, the use of brass shim stock is suggested. Select a piece of shim stock

.001 inch thinner than the maximum clearance specified in Table 5 in the appendix. Cut the shim stock to slightly less than the width of the bearing being checked and about ¼ inch wide. Use an oil stone on the shim so that there are no rough or turned over edges to damage the bearing material. Remove the bearing cap and coat the shim with light oil. Place the shim on the crankshaft journal or in the bearing shell and install the cap and tightening to proper tension. The other main bearing caps must be loosened when making this check. Rotate the shaft through a 2 inch arc—1 inch each way. If a definite drag is felt, the clearance is correct. Too little drag means bearing clearance is too great or if the shaft cannot be rotated, the clearance is insufficient.

The crankshaft end thrust is absorbed by the flanges on the rear main bearing shells. If the shaft end play is not .002 inch to .006 inch measured as shown, these flanges may be dressed down by rubbing them on a piece of emery cloth laid over a surface plate.

The Oil Pump Drive Cover and Idler Gear

Slide the oil pump idler gear on the shaft, and place the oil pump drive cover, gear, and gasket in place on the cylinder block. Install the cap screws and locking wire.

The Flywheel

Place the flywheel in position on the crankshaft. Install the dowel retainer and cap screws on the flywheel, and tighten the cap screws to from 65 to 70 foot-pounds. Check the flywheel for run-out with a dial indicator. If the run-out is more than 0.005 inch, make certain no foreign matter or burrs are between the flywheel and the crankshaft. Recheck the flywheel for run-out, and if there is still a run-out of more than 0.005 inch, take off the flywheel, turn it 180 degrees, and install it again. If there is still a run-out of more than 0.005 inch the flywheel must be replaced or resurfaced. Lock the cap screws with wire. Pack the pilot bearing with a short fiber sodium soap grease having a melting point of not less than 300 degrees F.

Note: If the engine is equipped with liquamatic drive, position the fluid coupling on the crankshaft flange and install the caps and screws.

The Clutch Disk and Pressure Plate

Block the three clutch levers down as shown in Fig. 2-10. Hold the clutch disk in place and install either a clutch shaft or a

Fig. 10-2. Checking main bearing clearance.

clutch pilot tool into the clutch disk and pilot bearing. Place the clutch pressure plate on the flywheel, and install and tighten the cap screws and lock washers. Remove the blocks that hold the clutch release levers. Remove the clutch pilot tool.

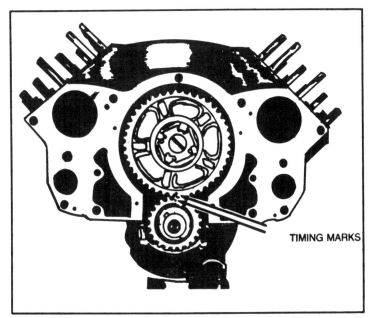

TIMING MARKS

Fig. 10-3. Location of engine timing marks.

The Camshaft Assembly

Lubricate and slide the camshaft into the cylinder block, making sure the timing mark on the camshaft gear is in line with the timing mark on the crankshaft gear (Fig. 10-3.)

The Push Rods (Lifters) and Valve Assemblies

Lubricate and place a push rod in each push rod bore. If any of the push rods are tight in the bore, select a push rod which will slip into the bore by its own weight.

Note: If the valves are new, do not use push rods which have been refaced.

Turn the camshaft until No. 1 push rod is resting on the heel of the cam. Install No. 1 valve assembly in No. 1 valve port. Pull the valve guide bushing down with a bar type valve lifter, and insert a valve guide bushing retainer in the bushing. Upon removal of the valve lifter, be sure that the retainer is seated in the slot of both halves of the valve guide bushing.

Check the clearance between the push rod and the end of the valve stem with a thickness gauge. If the clearance is more than 0.012-inch for the intake or 0.016-inch for the exhaust, select a longer valve or reface the valve or valve seat to lower the valve. If the clearance is less than 0.010-inch for the intake or 0.014-inch for the exhaust, select a shorter valve or grind the lower end of the stem until a clearance of 0.010-inch to 0.012 -inch for the intake and 0.014-to 0.016-inch for the exhaust is established. Repeat the above operation for each valve.

The Piston Rings

Select rings comparable in size to the pistons being used. Figure 10-4 shows the various types used in these motors. Data Table 4, in the Appendix, lists the various measurements we will be working with.

Slip the ring in the cylinder bore and press the ring down into the cylinder bore about 2 inch, squaring it with the cylinder; check the gap between the ends of the ring with a feeler gauge. If the gap between the rings is less than the specifications called for remove the ring and try another fit. Continue this operation until each ring is fit separately to the cylinder in which it is going to be used.

Carefully remove all foreign matter from the ring grooves, and inspect the grooves carefully for any burrs or nicks that might cause the ring(s) to hang up.

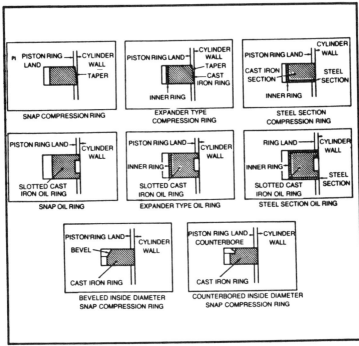

Fig. 10-4. Types of piston rings.

Slip the outer surface of the compression ring into the piston ring groove and roll the ring entirely around the groove to make sure that the ring is free and does not bind in the groove at any point. If binding occurs, the cause should be determined and removed by carefully dressing with a fine cut file. If, however the binding is caused by a distorted ring, replace it with a new one. Each ring and groove should be checked as described.

Proper clearance of the piston ring in its piston ring groove is very important in maintaining engine performance and in preventing excessive oil consumption. When fitting new rings, the clearance between the top and bottom surfaces of the grooves should be inspected.

When installing your rings, always use a piston ring expander. The compression rings are the uppers while the oil rings are the lowers. Be sure that when fully compressed into the piston grooves the rings do not butt together before they are flush with the piston side. Once all the rings are installed on a piston, rotate the rings so that the gaps are not aligned; this will give a better seal, helping to insure against blow-by.

Assembly of the piston to rod was covered in Chapter 4. Refer to the sections **Piston and Con Rod Disassembly** and **Piston Pins and Bore** for your final piston to rod assembly sequence and specifications.

The Connecting Rod and Piston Assemblies

The connecting rod bearing total clearance between the shaft and the insert, and between the insert and the connecting rod bore on a new engine, is 0.0015-inch to 0.0035-inch. Parts must be selected to obtain a total clearance not to exceed 0.005-inch when old parts are used. If previous inspection indicates any wear under the specified limits on the crankpin journals, connecting rod bearings or the bore of the connecting rod, a combination of these worn parts may exceed the allowable clearance of 0.005-inch. Follow whichever of the following conditions that apply:

☐ **Crankpin and Rod within Manufacturing Limits.** If the crankpin and connecting rod bores within manufacturing limits, use standard size bearings.

☐ **Crankpins Undersize, Rod Bores within Wear Limits**. If the crankpins have been reground or are worn more than 0.0015-inch undersize and the connecting rod bores are within the wear limits, use a bearing with its outside diameter standard and its inside diameter undersize as required.

☐ **Crankpin Bore Oversize.** If a connecting rod bore has been reground oversize, select a bearing having the correct inside diameter of the crankpin and the correct outside diameter for the oversize connecting rod bore. It will be necessary to use this rod in the same crankpin with a second rod of the same oversize.

Note: Before installation, select the piston assemblies for each cylinder as outlined in Chapter 4.

Oil The Piston Rings

Lubricate the two halves of No. 1 connecting rod bearing and place them on the crankpin. If the old bearings are used, install them on the original crankpin from which they were removed. Place the piston and connecting rod assembly marked R1 in No. 1 cylinder in the right-hand bank with the connecting rod and piston assembly number facing toward the front of the engine. Use a piston ring compressor on the piston rings and tap the piston down into the cylinder with the end of a hammer handle. Position the connecting rod on the crankpin and install the bearing cap on the connecting rod, making sure the number on the connecting rod is facing toward the front of the engine. Install, but do not completely tighten, the nuts. Place the piston and connecting rod assembly marked L-1 in No. 1 cylinder in the left-hand bank in the same manner and attach the connecting rod to the crankpin. Repeat the above operation, installing the remaining connecting rod and piston assemblies. Tighten the connecting rod nuts to from 35 to 40 foot-pounds, and install cotter pins in each connecting rod.

Note: If self-locking nuts are used on the connecting rod studs, they are to be tightened to from 40 to 45 foot-pounds.

The Oil Pump

Place the oil pump assembly in position in the cylinder block, and install the cap screw and locking wire.

The Cylinder Front Cover

Note: Soak new cylinder front cover oil seals (part 6700) in the oil for approximately two hours before installation.

Install the oil seal in the recess provided in the cylinder front cover, making sure the ends of the seal protrude evenly at both ends. Place the cover and gasket in place on the cylinder block (Fig. 10-5). Install the cap screws and lock washers.

The Crankshaft Pulley and Starting Crank Ratchet

Install the crankshaft pulley Woodruff key in the crankshaft. Tap the pulley onto the crankshaft and install the flat washer and starting crank ratchet on the crankshaft.

The Oil Pan

Note: Soak new oil pan front seals in oil for approximately two hours before installation.

Install the oil pan front seal in the oil pan, making sure the ends of the seal protrude evenly at both ends. Coat the bottom machined surface of the cylinder block with grease, and set the oil pan gaskets in place. Install the oil pan rear cork packing in the recess provided in the rear main bearing cap. Place the oil pan in position on the cylinder block and install the cap screws and lock washers; tighten to from 15 to 18 foot-pounds.

The Water Pumps

Hold a water pump and gasket in position on the cylinder block. Install the cap screws and lock washers in the pump, and tighten the cap screws.

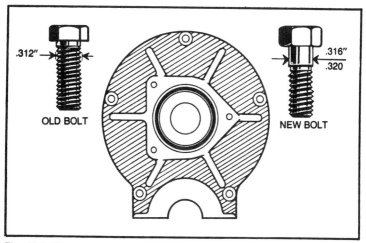

Fig. 10-5. Engine front cover. Later engines are equipped with a front cover having a .324″ reamed hole. The dowel type cap bolts shown should be used with this cover, which accurately locates the front cover, which in turn assures the proper distributor location. Note: Similar dowel screws are used in locating the distributor.

Note: One of the cap screws is installed through the pump inlet opening on '37-'53 blocks. Repeat the above procedure on the other water pump.

The Cylinder Heads

Cylinder blocks having a basic part number prefix 41A or 59A can be identified by oblong water passages on the gasket aluminum heads.

Unless the camshaft has been changed from original design, there is usually no need to check valve-to-head clearance, although it is never a bad idea.

To check this clearance, lay thin strips of modeling clay atop the valves. Bolt and torque the head in place, using the old headgasket. Rotate the motor at least two times, then remove the heads. If the clay measures less than .060″ where it has been pressed between the valve and head, more clearance will be necessary. If more clearance is needed, the head chamber above the valves must be fly cut the necessary amount.

The Intake Manifold

Place the intake manifold gasket on the cylinder block. Making sure there is no foreign matter in the valve chamber, place the intake manifold on the cylinder block, and install the lock washers and cap screws, and tighten all the cap screws except the four cap screws that hold the two spark plug wire conduits to the intake manifold.

The Exhaust Manifolds

Place the right and left exhaust manifolds and gaskets in position on the sides of the cylinder block. Coat the threads of the exhaust manifold cap screws with graphite, and install the lock washers and cap screws.

Installation of Accessories

The following instructions are based on the assumption that the various accessories are in good working order. Overhaul or repair the various accessories before installation if required.

☐ Install the Starting Motor. Hold the starting motor in position on the flywheel housing. Install and tighten the cap screws that hold the starting motor bracket to the oil pan.

☐ Install the Spark Plugs. Clean the spark plugs and set the gaps to 0.025-inch. Install the eight spark plugs and gaskets in the cylinder heads and carefully tighten the spark plugs to from 24 to 28 foot-pounds.

☐ Install the Fuel Pump and Adapter. Install the fuel pump push rod in the cylinder block.

Warning: Replace the push rod if its length is worn to less than 8.870 inches.

Install the fuel pump adapter gasket on the intake manifold. Install the fuel pump and adapter on the intake manifold. Install the lock washers and nuts on the fuel pump, and tighten the nuts.

☐ Install the Carburetor. Place the carburetor gasket in position on the intake manifold, and install the carburetor on the intake manifold. Install the lock washers and nuts on the carburetor studs, and tighten the nuts. Connect the fuel line to the carburetor and fuel pump.

☐ Install the Generator. Place the generator on the intake manifold and install, but do not tighten the generator mounting bracket nut.

☐ Install the Belts and Fan. To install the bracket mounted fan, position the generator belt on the crankshaft, water pump, and generator pulleys. Place the fan and bracket assembly in position on the generator mounting bracket, and install, but do not tighten, the two cap screws which hold the fan and bracket assembly to the generator mounting bracket. Install the fan belt on the fan and crankshaft pulleys. Raise the gererator up until a total movement of 1 inch of the belt is possible at a point midway between the generator and water pump pulleys. Tighten the generator support bracket nut. Raise the fan and bracket assembly up until a total movement of 1 inch of the belt is possible at a point midway between the fan pulley and the crankshaft pulley. Tighten the fan bracket assembly cap screws.

A new motor can wear more in the first minute after it is started up than in the next six months, if it has a prolonged dry start. Special molybdenum disulphide oil additives with moisture absorbing oxidizers are available (STP, Iskenderian cam lube, etc.) as break-in additives. These additives, and others like them, help to deter internal engine rust and oil dilution in the crankcase. A liberal coating of the critical internal parts of your motor during the final assembly stage with one of these additives will aid the first few critical minutes of the engine's initial fire-up. As a break-in additive, it will help to reduce friction, prevent metal to metal contact and help to assure effective lubrication during the break-in period.

An oil pump primed with an MD additive will start oil circulating almost instantly, whereas one that has to start dry may take several precious seconds before it can suck oil up from the pan and distribute it through the motor's oiling system.

Chapter 11
Engine Start-up and Break-in

Now that everything is back together, comes the final moment of truth: Will she or won't she. If you've been methodical and meticulous throughout this project, she will. In fact, with proper care and maintenance, she'll provide you with years of service and driving pleasure. But before you try to fire that newly rebuilt motor, make sure you've covered all the bases. Once you feel confident that everything is in order, fire away. If trouble should occur in a given area, refer to that part of the book which deals with the troublesome system component. At any rate, good luck with the initial starting, break-in period and shake-down cruise.

Once Again . . . Oil

In several areas throughout this project, I have talked about the importance of oil. The importance of adequate lubrication in a motor cannot be over emphasized. Heed the warnings: quality oil is cheap insurance. Any major brand, good quality 40 wt. S.A.E. oil is fine for your motor. Personally, I prefer Pennzoil or Quaker State, which is neither here nor there. Just use a good quality oil. Your motor will require 4 quarts plus 1 quart if an oil filter is used.

Timing the Engine

Setting the valve tappets to camshaft clearance (lash) and coordinating the distributor firing sequence with the compression stroke of the cylinder (distributor timing,) is a somewhat wordy way of describing what is known as Engine Timing. Setting of the valve tappet clearance (lash), was covered in Chapter 10, Engine Reassembly. Setting of the distributor timing was covered in Chapter 8, The Ignition System.

Once the motor is assembled, back in the car, and ready to be started for the first time—and assuming you have the correct valve lash setting—there are two major areas of difficulty that most of us get into.

The first is having the motor fire on the exhaust stroke, rather than on the compression stroke. When this happens, recheck your procedure to see that the distributor is not out of phase with the cylinders.

The second biggest problem is having the spark plug wires crossed. I've made this mistake more often than I care to talk about, but it is a common one and easy to rectify. When you suspect a crossed wire or when you're installing your wires initially, start from scratch:

☐ Remove the No. 1 spark plug and locate the compression stroke as described in Chapter 8, The Ignition System. Remove the distributor cap(s) and observe the rotor as you slowly rotate the engine. When the piston is at the top of its compression stroke, the rotor should be on or closely approaching a terminal relative to the distributor cap. Make a chalk mark on the outside of the distributor housing, parallel to the rotor pick up point. Reinstall the cap(s). The chalk mark will be in alignment with one of the spark plug wire terminal towers of the distributor cap. Make a permanent mark on this tower indicating No. 1, if it is not already embossed onto the cap by the manfacturer.

☐ Remove the distributor cap(s) and, in a clockwise pattern, mark the engines firing order (15486372) on the cap; refer to the engine electrical schematics in Chapter 9.

☐ Mark your wires on each end 1 thru 8 and reinstall the distributor cap. Route your wires in numerical order from the cap to the corresponding cylinders, spark plugs. Cylinders 1-2-3-4 are on the right bank of the motor (passenger side); cylinders 5-6-7-8 are on the left bank (driver's side). Cylinders 1 and 5 are forward and cylinders 4 and 8 are at the rear.

☐ You should now have everything in its proper perspective and eliminated the common problems. If trouble still persists, refer to Chapter 12, Troubleshooting, or the engine electrical schematics in Chapter 9, The Electrical System.

The Rebuilt Motor

There's one thing certain about breaking in a rebuilt motor; if you ask a dozen different people the right way to do it, you'll get a dozen different answers. Everyone seems to have a secret formula for breaking in a motor and is convinced that it is the only right way to do it. There seems to be one common element in all of them, though; everyone will agree that the most important factor in breaking in any engine is the seating of the piston rings. Getting a good sealing surface between the rings and the cylinder wall is what determines whether or not the engine will burn oil and also whether or not it will make horsepower. The best way to establish a good ring seal is by running the engine at varying rpm during the first few hundred miles.

Be sure to spend some time in deceleration as well as acceleration during the break-in. The ring seal is governed by the gas pressure exerted on the rings and that pressure is determined by engine speed. Running at a constant speed will generate a balanced pressure and may lead to polishing of the cylinder walls. Therefore, continually varying the speed during break in is of utmost importance. The acceleration/deceleration cycle is necessary because this will load both sides of the rings against the walls; since the thrust surfaces are loaded during acceleration and the opposite surfaces are loaded during deceleration.

As soon as the motor is started, it should be set at high idle, which will allow oil to circulate through the engine oil galleries; oil pressure is very important, so keep an eye on your oil pressure gauge. Also, keep a very close watch on the engine temperature gauge to avoid any chance of overheating, which may distort the internal components at this critical time in the engine's new life. After the engine has reached operating temperature and runs for 20 to 30 minutes, recheck the ignition timing and adjust the carburetor for correct idle speed. Shut the engine off and retorque the head and manifold bolts. Recheck oil and coolant levels.

Don't take any long trips and don't run the engine at extreme rpm. After the first couple of hundred miles, change the oil and oil filter. It's also a good idea to spend some time checking the engine compartment for loose belts, hoses, electrical connections and so on. Once the 200-mile mark is reached, you may then go ahead and run the engine the way you plan to drive it. If it's not broken in by the time you reach the 500-mile mark, it probably never will be.

Chapter 12
Troubleshooting Your Flathead

This chapter is intended to provide guiding information so you can quickly locate and eliminate the source of trouble in a particular system. Troubleshooting, or diagnosis as it is sometimes called, is the establishment of facts first, to find out whether or not a suspected trouble actually exists; then, to track down the source in a logical manner. The troubles which you may encounter will appear as symptoms under descriptive headings which will aid you in locating the procedure you need.

Engine Does Not Develop Full Power

If the engine does not develop full power, has a low top speed, or slow acceleration, it is advisable to perform a complete engine tune-up. In most cases, this will correct the trouble and will eliminate many of the following procedures. If the tune-up was performed by someone else, or if it did not correct the trouble, proceed as follows:

Combustion. Have a combustion analysis made to determine if the air-fuel ratio is correct. If the air-fuel ratio is too rich or too lean, clean and adjust the carburetor.

Cylinder Compression. Check the compression of each cylinder. If the compression checks above normal, excessive carbon deposits are indicated. Remove the cylinder head and clean out the carbon.

If the compression checks below normal, the rings are worn or the valves are not seating properly. Squirt a small quantity of light engine oil in the cylinder and retest the compression. If the compression is now normal, the rings, pistons, or cylinder walls are worn and must be repaired. If the compression is still low, the valves are leaking. Regrind the valves and seats (or replace warped valves).

Ignition. Remove each spark plug wire in turn and hold it 3/16-inch away from the cylinder head. If the spark jumps this gap consistently (engine idling) at each plug, the system can be considered to be operating properly.

If the spark is not consistent at any wire or several wires, the trouble lies in the ignition system. See "Satisfactory Spark from Some But Not All Spark Plug Wires" below.

Fuel Pump. Test the fuel pump pressure and vacuum. If these are satisfactory, make a volume test of the pump. If the pump will not fill a pint measure in 45 seconds, check the fuel lines for obstructions. Clean out the lines if possible. If not, replace defective lines. If the lines are clear, repair or replace the fuel pump.

Additional Possible Causes. If the trouble still exists after the above tests and corrections, check for the following additional causes: Dragging brakes, excessive exhaust back pressure (plugged lines, etc.), camshaft out of time, too little or too much valve clearance, misalignment of front wheels, and non-standard equipment (tire size, rear axle ratio, speedometer gear, etc.).

Engine Runs Unevenly or Backfires

Make certain the choke is operating properly and the engine is sufficiently warmed up.

Spark Plug Wires. Check for correct position of the wires in the distributor cap. The firing order for eight-cylinder engines is 1-5-4-8-6-3-7-2 and for six-cylinder engines is 1-5-3-6-2-4. Occasionally, the spark may be "cross firing" from one wire to another due to worn insulation.

Distributor Cap. Check for a cracked or shorted distributor cap. Replace the cap if it is cracked or shorted.

Fuel System. Check for leaks in the carburetor, fuel pump and lines. Make sure the vent in the gas tank cap is unrestricted. Blow out blocked fuel lines.

Engine Misfires at High Speed

The most probable cause of trouble is poor ignition. Another possible cause is an improperly adjusted carburetor. If ignition and carburetor are in good working order, check for sticking valves.

Spark Plugs. Clean and adjust spark plugs. File the electrodes with an ignition file. Make sure the tips and edges of the electrodes are sharp and square.

Distributor. Check the point gap and reset if necessary. Be sure the spark is timed correctly. Check the condenser and replace if defective.

Fuel System. Be sure the carburetor float level is correct. Check the fuel pump and fuel lines for restrictions.

Engine Cranks But Will Not Start

This trouble is often caused by temporary conditions which can be readily corrected.

Spark Plugs. Check the spark plug insulators for a dirty or wet surface. Wipe off any dirt or moisture.

Distributor Cap and Wiring. Wipe off dirt and moisture from the distributor cap and high tension wires.

For Vapor Lock Or Flooding. If the engine is hot, a vapor lock may be blocking the flow of fuel to the carburetor. If the engine has been choked too much or if the foot throttle has been over-manipulated, the engine may be flooded.

In either case, push the choke button all the way in and crank the engine for several revolutions with the foot throttle wide open.

Insufficient Choking. If the engine is extremely cold, it may be necessary to pull the choke button all the way out to get the engine to start.

Fuel At Carburetor. Remove the air cleaner and operate the throttle linkage. Look for a spray of gas inside the carburetor from the accelerating pump each time the throttle is operated. If there is no spray, check for a faulty fuel pump or plugged fuel lines.

Ignition. Remove any spark plug wire and hold the terminal 3/16-inch from the cylinder head. Crank the engine (ignition on) and observe if a spark jumps the gap regularly. If there is no spark, refer to "i. No Spark At Any Spark Plug Wire" below.

Engine Misfires on Acceleration or Hard Pull

The most probable cause for engine "miss" is the ignition. Other possible causes are insufficient fuel and sticking valves.

Spark. Remove each spark plug wire in turn and hold it 3/16-inch from the cylinder head. If a spark jumps this gap regularly proceed to "Clean and Regap Spark Plugs." If the spark does not jump regularly or there is no spark from any plug wire, see "No Spark At Any Spark Plug Wire" below.

Spark Plugs. Remove the plugs and clean the insulators. File the electrodes and reset the gap.

Accelerating Pump. Set the pump link in the proper hole for the prevailing temperature. The inner hole (shortest pump stroke) is for very hot temperatures, the center hole for normal temperatures, and the outer hole for extremely cold temperatures.

Float Level. Check the float level to see if it is too low. A low fuel level in the float bowl can "starve" the carburetor when the engine is running at high speed.

Sticking Valves. If the engine still misses after the above procedures, listen for abnormally noisy valves. The noise can indicate sluggish valve action. Clean the valve stems and guides.

Fuel Not Reaching Carburetor

A clogged or broken fuel line or a faulty fuel pump is the most common cause of this trouble. Check the supply of fuel in the fuel tank. Make sure the tank vent is open.

Fuel Line. Remove the flexible tube in the fuel pump and replace the tube if it leaks air or if the fuel passage is obstructed. Remove the fuel tank filler cap and blow out the fuel line.

Fuel Pump. Remove the fuel line between the fuel pump and the carburetor. Blow through the line to make sure it is not clogged. With the ignition switch OFF, crank the engine with the starter. If a free flow of fuel is not evident, the fuel pump is faulty and must be repaired or replaced. If the fuel pump and the fuel line are found satisfactory, remove the carburetor and clean the carburetor float valve mechanism.

Carburetor Floods

Flooding is caused by a sticking choke, high fuel pump pressure, carburetor float set too high, or dirt in the fuel inlet valve. On hot days, fuel may "percolate" from the carburetor bowl into the intake manifold when the engine is stopped, causing flooding of the manifold.

In addition to the engine running unevenly, a strong odor of gasoline usually is present when the engine is flooded. If the flooding is due merely to overchoking, open the throttle wide and crank the engine to exhaust rich gases.

Carburetor Choke Action. Remove the air cleaner, operate the choke and observe if the carburetor choke plate opens freely. If the choke action is faulty, make the necessary corrections.

Fuel Pump Pressure. Test the fuel pump pressure with the engine running at idle speed. If the pressure is higher than normal, make the necessary repairs or replacements.

Remove and Disassemble Carburetor. Remove and disassemble the carburetor and clean all parts. Examine the float for leaks and check the condition of the float needle valve and seat. Make repairs as required and set the float level. Install the carburetor.

Fuel Mixture Too Lean

This indicates an insufficient supply of fuel passing through the carburetor for the volume of air drawn by the engine. Obstructed fuel lines, low fuel pump pressure, or a leaking intake manifold may be the cause.

Fuel Tank and Lines. Make sure the fuel pump and connections are not leaking. Make sure the fuel vent is open. Remove the flexible line at the intake side of the fuel pump and replace the line if there is any indication of leakage. Remove the fuel tank cap and then blow compressed air back through the fuel line.

Fuel Pump. Check fuel pump pressure and vacuum. If the fuel pump pressure and vacuum are satisfactory, check the fuel pump capacity. If the fuel pump is not operating within limits, make the necessary repairs or replacements.

Carburetor. Remove, disassemble, and clean the carburetor. Make all necessary repairs. Set the float level. Install the accelerating pump link in the proper hole for the prevailing temperature. Make sure the throttle linkage permits full opening of the throttle plate.

No Spark at Any Spark Plug Wire

This indicates trouble in the primary circuit or in the high tension wire between the coil and the distributor. Schematic drawings of the ignition circuits are shown in Fig. 12-1. The numbers appearing in these drawings establish the locations of units in the circuits.

Coil to Distributor High Tension Wire. Replace the coil-to-distributor high tension wire if the insulation is worn or damaged at any point. Make sure the terminal (16) is soldered to the wire and is firmly seated in the coil terminal socket. Make sure the coil to distributor primary wire is making good contact at both ends.

Jumper Between Battery and Coil. Connect a jumper wire between the battery negative terminal (14) and the battery terminal of the ignition coil (7). Turn the ignition switch off. Crank the engine with the starter, and follow the procedure under the heading that agrees with your observation.

If the engine starts, the trouble is in the primary circuit from the negative (hot) side of the starter relay to the battery terminal of the ignition coil (13 to 7). Note: Do not run the engine for more than five minutes with the wires connected in this manner. Momentarily disconnect the lead from the battery (14) to stop the engine. Working from the coil toward the battery, contact the ammeter jumper lead consecutively to each of the primary circuit terminals (7 to 13) until the engine starts. The faulty part of the circuit is between the terminal where the engine would not start and the terminal where it will start. Clean corroded terminals, tighten terminals, and repair or replace parts at fault.

If the engine does not start, the trouble is in the primary circuit from the battery terminal (7) of the coil to the ground side (1) of the distributor points. Replace or adjust distributor points or repair the primary circuit contact (3).

If the primary circuit is in good condition, replace the condenser. Remove the high tension lead from the distributor cap center terminal and hold the end of the lead 3/16-inch from the cylinder head while cranking the engine (ignition on). If there is no spark, replace the coil. If there is a spark, check the distributor cap and rotor for cracks or for spark tracks. Clean or replace the cap.

Satisfactory Spark from Some but Not All Spark Plug Wires

The trouble lies within the distributor cap or the remainder of the secondary circuit to the spark plugs. The fact that a satisfactory spark is obtained from some spark plug wires eliminates from consideration those factors that affect equally the output of all the spark plug wires (the primary circuit).

Spark Plug Wires. Replace spark plug wires if the insulation is damaged. Make sure all spark plug wires are soldered to their terminals. Make sure the spark plug wire terminals and the terminal sockets are free from corrosion and the wires are firmly seated in the distributor cap. If the above procedure has not corrected the trouble, proceed.

Distributor Cap. Remove the distributor cap and clean the cap with lacquer thinner. Replace the distributor cap if it is burned or has carbon tracks. Make sure the spark plug wires seat firmly in the sockets.

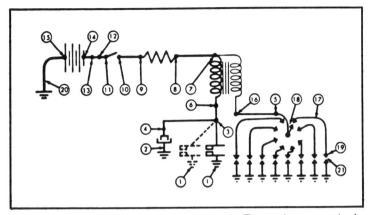

Fig. 12-1. Schematic drawing of the ignition circuits. The numbers appearing in this drawing establish the location of units in the circuits as referred to in the text.

Intermittent or Weak Spark at all Spark Plug Wires

The trouble is in the primary circuit or in the coil to distributor high tension lead.

Connections. Tighten all connections in the primary circuit, including both terminals of the condenser (2 and 4) and both ends of both battery cables (13, 14, 15, and 20).

Make sure the coil-to-distributor high tension wire terminal (16) is soldered to the wire and seated all the way into the high tension terminal of the coil. If the trouble still exists, proceed.

Distributor Points. Replace or adjust the distributor points, if required. Make sure the breaker arm is not binding and that the spring tension is correct. Reset the timing, and again test the quality of the spark.

Condenser or Coil. Replace the condenser. If the spark does not improve, a new coil must be installed.

Engine Overheats

The engine will overheat when there is insufficient transfer of heat to the coolant and to the air passing over the engine.

External Leakage. Fill the cooling system and idle the engine. Inspect all hose and hose connections for leakage. Tighten the connections or replace hose as required. Inspect the radiator cap for tightness and note the condition of the gasket. If leakage is observed at the cylinder head gaskets, replace the gaskets (including remove carbon). Inspect the radiator for leakage, and repair or replace if required. Rust spots or wet spots on the radiator core are an indication of radiator leakage.

Fan and Water Pump Belts. Adjust or replace the fan and water pump belts if required.

Exhaust System. Inspect the exhaust pipes, muffler, and tail pipe for evidence of dents, kinks, collapse, or restrictions of any kind. Make any necessary correction.

Radiator Core. If the air flow through the radiator is restricted (insects, leaves, grease, dirt, etc.), clean the fins and air passages.

Ignition. Time the ignition and check the spark advance.

Radiator Hose. Inspect the radiator hose and replace any hose that has become soft or collapsed.

Thermostats. Remove each thermostat and place it in hot water. Thermostats should operate according to the temperatures given in its specifications. High temperature thermostats used in connection with hot water heaters open at slightly higher temperature. Replace any faulty thermostats.

Cooling System. Flush the cooling system with Ford cooling system cleaner.

Internal Leakage or Deposits. Drain the oil from the engine oil pan and check for water in the oil. If you find an abnormal amount of water in the oil, remove the spark plugs and see if water is present at the plug holes.

With the engine cold, fill the radiator to the top. Remove the fan belt so that the water pumps will not operate. Run the engine at fast idle for 60 seconds. If water runs out of the radiator filler pipe or overflow, or if bubbles come to the surface of the water in the radiator, leakage exists between one or more of the cylinders and the cooling system. Tighten the cylinder heads. If leakage is still evident, remove the cylinder head and inspect for broken gaskets or a cracked cylinder block. Replace broken gaskets or make necessary corrections in case of a cracked block. Inspect the water openings in the cylinder heads and cylinder block for excessive lime deposits. If excessive lime deposits are present, the cylinder block and heads must be replaced as flushing will not remove lime deposits.

Note: Excessive deposits of lime are the result of using water with a high mineral content in the cooling system. Always use soft water if possible.

Battery Low in Charge

Indications pointing to this symptom are: Slow cranking, hard starting, and headlights dim at idle engine speed. Causes of this symptom are: The generator belt is worn or loose and slipping over the generator pulley. The battery may be in such a poor condition that it will not hold or take a charge. The generator may not be producing its rated output. Regulator units may be out of adjustment or in need of repair. Excessive resistance may exist in the generator-to-battery circuit or in the battery-to-ground circuit. The car may not be driven enough to keep the battery charged.

Before starting the following procedure, to confirm the symptom, check the state of charge of the battery. Likewise, check the generator belt adjustment and condition. If the belt is broken or too loose, this might be all that is needed.

Generating System (Two Brush Generator). Disconnect the "Field" wire at the voltage regulator and connect this wire to the "ARM" terminal of the regulator. Start the engine and increase engine speed. If the charge indicator now shows a charge, the generator is in operating condition.

If there is no charge indicated, reduce the engine speed to idle, remove the "BAT" lead from the regulator, and connect the lead to the "ARM" terminal. Speed up the engine and observe the charge indicator. If the generator now charges, the cutout unit of the regulator is defective. Repair or replace the regulator. If no charge is indicated, see "Generator Output Low."

Reduce the engine speed to idle, remove the regulator leads and then stop the engine.

Caution: Do not stop the engine with the leads still connected. You may burn out the generator.

Generating System (Three Brush Generator). Remove the cut-out cover. Adjust the third brush on the generator for maximum output. Start the engine and slowly increase the engine speed. Observe the cut-out to see if it closes. If the cut-out does not close, connect a jumper wire from the cut-out "ARM" terminal to the "BAT" terminal. If a charge is now indicated, the cut-out will have to be adjusted, repaired or replaced. If no charging rate is indicated, see "Generator Output Low."

Generator Output Low

Generator output can be low due to dirty commutator, open or short circuit in the field, armature, brushes, or brush holders, or the brushes can be worn too short or sticking in the brush holder and not making good contact on the commutator.

Substitute Ground Jumper. Connect a jumper wire from the generator "GRD" (ground) terminal to the battery ground post.

Run the engine and see if the charging rate is increased. If the rate is increased, the ground circuit is defective. Clean the cable terminals. Remove the generator from its mounting bracket and clean the mating surfaces. Also, clean paint, grease, or dirt from the mating surfaces of the bracket slide and the engine block.

Commutator. If the charging rate is not increased, squirt carbon tetrachloride on the generator commutator. If the charging rate is now increased, the commutator is dirty and should be cleaned and polished. If the charging rate does not increase, the generator is defective and it must be repaired or replaced.

High Charging Rate

Indications of this symptom are: Generator, lights, or radio tubes burn out prematurely, the battery requires too frequent refilling, and ignition contacts are burned. The most common cause of these troubles is high voltage, and the first step of troubleshooting is to correct possible high voltage regulation. In cases where the two brush generator itself burns out, in addition to the high voltage, a high setting of the current limiter could account for the failure.

Adjust Voltage Regulation (Two Brush Generator). Have the voltage regulator adjusted by your Ford dealer or by a competent electric service shop. The voltage regulator is a complex device and adjustments or repair require accurate test equipment technical knowledge about the regulator.

Voltage (Three Brush Generator). To decrease the voltage of the three brush generator, remove the cover band, loosen the brush plate lock screw, and slide the brush in a direction opposite to the rotation of the armature.

Starter Does Not Crank Engine

Several causes may result in this symptom. The battery may be discharged, cables may be broken, the starter switch may be defective, or the starter motor may be inoperative.

Before starting the procedure for either type system, check the state of charge of the battery. If the battery is discharged it should be recharged, check for capacity, and replace if the capacity test shows it is worn out. If the battery is OK, proceed with the following tests.

Remote Push Button System. This system has a starter relay to close the starter circuit and a push button mounted on the dash panel to actuate the relay. Press the starter button and listen for a click at the relay. If the relay clicks, proceed to check by-pass relay.

If the relay does not click, ground the starter button wire terminal at the relay. If the engine cranks, the starter button or the wire connecting the button and relay is defective; check the connecting wire for breaks. If the wire is OK, replace the button.

If the engine does not crank and the starter relay does not click, replace the relay.

Connect a starter cable jumper to one of the heavy terminals on the relay. Touch the wire to the other heavy terminal. If the engine cranks, replace the relay.

If the engine does not crank, inspect the cables and clean the connection. If the engine still will not crank and the cables are in good condition with all connections clean and tight, the starter motor is defective and must be repaired or replaced.

Foot Switch Starting System. This system has a direct acting foot switch in the starter circuit. Connect a heavy jumper from one terminal of the relay to the other. If the engine now cranks, replace the foot switch.

If the engine does not crank, inspect the cables and clean the connections. If the engine still will not crank and the cables are in good condition with all connections clean and tight, the starter motor is defective and must be repaired or replaced.

Starter Spins But Will Not Crank Engine

In either type system, the starter drive is dirty or worn and is sticking on the starter shaft. Remove the starter motor and clean the starter drive by washing it in kerosene. Replace worn or damaged parts. Install the starter. Note: Do not oil the starter drive. It should work freely when cleaned in kerosene.

Engine Cranks Slowly

The major causes of this symptom are: too heavy grade of engine oil, battery discharged, excessive resistance in the circuit, or defective starter motor.

For either type starting system, check the grade of engine oil. If it is too heavy for the prevailing temperatures, drain and refill the sump with lighter grade oil. Check the state of charge of the battery. Recharge the battery if required. If the battery is worn out, replace it. If the battery checks OK, proceed below.

Cables and Connections. Remove the heavy starter cables and check their condition. Replace any worn or broken cables.

Clean all terminal connections and replace the cables. If the starter still turn slowly, proceed with 2.

By-Pass Starter Switch. Connect a jumper to the heavy terminals of the starter switch. If the engine now cranks at normal speed, the switch is defective. If the engine still cranks slowly, the starting motor is defective. Repair or replace the starting motor. Note: If the motor is known to be in good condition, the engine has excessive friction (too tight bearings, pistons, etc.).

Headlights Flicker from Bright to Dim

If all lights flicker from bright to very dim, the overload circuit breaker is operating as a result of a grounded or shorted wire in that particular circuit. Set the headlight switch to headlight position. Observe the reaction as you switch from high to low beam with the beam control switch.

If the lights flicker when on low beam, the short is in that circuit, likewise, the short is in the high beam circuit if the high beam lights are on.

Upper Beam Only Flickers. If the lights flicker only when the beam control switch is in the upper beam position, the "short" is in the upper beam circuit from the beam control switch to the headlights.

Lower Beam Only Flickers. If the lights flicker only when the beam control switch is in the lower beam position, the "short" is in the lower beam circuit from the beam control switch to the headlights.

Both Beams Flicker. If the lights flicker on both high or low beam, set the headlight switch to the parking light position. If the lights still flicker, a short exists in the taillight circuit. If the lights no longer flicker, a short exists between the headlight switch and the beam control switch.

Individual Lights Do Not Light

When one or several lights do not light and other lights do, the fault usually is in the bulb itself. However, some light bulbs are easily replaced while others present more difficulty. The ease with which the particular bulb can be replaced determines the order of procedure.

Bulb Readily Accessible. Replace the bulb. If this does not correct the trouble, proceed.

Bulb Difficult to Replace. Turn the lights on. Disconnect the wire at the bullet connector nearest to the bulb and momentarily ground the "hot" wire. If a spark occurs, connect the wire and replace the bulb or any wiring that runs from that point to the bulb. If no spark occurs, an open circuit exists between the point that was grounded and the light switch. Make necessary repairs.

One Or More Lights Burn Out Repeatedly

Clean and tighten all electrical connections in the circuit involved, including the battery cable connections. Test the generator voltage regulation and adjust or replace the regulator if required.

Horn Does Not Sound

If the horn does not sound when all of the wires are connected, press the horn button. If the horn still does not sound, disconnect the main feed wire at the horn relay and ground it momentarily and follow below, whichever applies.

Spark Occurs. If spark occurs, the wire can be considered satisfactory. Connect one end of a jumper wire to the main feed wire. Momentarily contact the other end of the jumper wire to each horn wire. If each horn sounds, replace the horn relay. If the horns do not sound, repair or replace the horns.

No Spark Occurs. If no spark occurs, an open circuit exists between the end of the wire that was grounded and the starter relay. Repair the wire.

Horn Sounds Continuously

To stop the horn from sounding, disconnect the horn button wire from the bullet connection at the lower end of the steering column or at the horn relay, whichever is more accessible. If the horn continues to sound, disconnect the horn wires from the horn relay.

If the horn stops sounding when the horn button wire is disconnected, repair or replace the horn button wire or horn button.

If the horn continues to sound after the horn button wire is disconnected, the trouble is in the relay or the wire between the bullet connection and relay. Repair the wire or replace the relay.

Chapter 13
Flathead Wrap-up

If it's ever been done to a flat motor, Tom Hutchinson has done it. Tom's been around these motors a long, long time. From Muroc to Bonneville to the Antique Nationals, from bone-stock to all-out, 160 mph competition, Tom has built (and raced) a lot of flatheads. Presently, he is owner/operator of a Ford flathead rebuilding service in La Puente, California.

A knowledgeable and dedicated flathead man, Tom supplies his region with parts and services. Prices are competitive for NOS, new manufacture and reproductions, while his stock of used parts go for unbelievable bargain prices. Tom has just recently completed a bone-stock '32 V8 and a fire-breathing ⅜ x ⅜ Ardun hemi conversion, both of which are for sale; several other used and rebuilt motors of various years are scattered throughout the shop.

Hemi, for those of you unfamiliar with the term, is short for "hemi-spherical combustion chamber." Very similar to Chrysler Corporation's Dodge, DeSoto, Plymouth and Chrysler Hemis, these conversions wipe out any existence of the L-head design and are for race-bred cars only. These English made Arduns develop brutal horsepower. They are very scarce and much sought after by hot rodders and racers.

Tom rebuilds and guarantees a flat motor for a very modest price. Rebuild includes a complete short block: magnaflux, bore, align bore, valve job, pistons, rings, bearings, rebuilt rods, crank and cam, plus all of the small miscellaneous items normally expected. The rebuild also includes one of Tom's own manufactured oil pumps and a special long lasting triple face valve job. The motor, when finished, is ready to run. The price is less heads. Manifolds, electrical, water pumps, etc., are extra or customer furnished. Sleeves are also extra if needed, but the cost is nominal. All motor parts are NOS, new manufacture or customer furnished. A good supply of blocks, cranks, cams, etc. are available if you cannot supply a core. ('32 V8's are a bit scarce). Tom can rebuild any year V8 from now until forever from his stock and suppliers. Four cylinder motors, transmission and rear-end work is also available. Tom is well known and respected in Southern California as a flathead man. He runs a complete and reputable service with customers worldwide.

Engine Balancing

This is a controversial subject and depending upon your school of thought, you may or may not agree that engine balancing is of any importance. I took a lot of lessons from the racers of these (and other) motors and I think the results in engine performance and longevity speak for themselves. In short, if the entire reciprocating assembly is balanced (by a competent engine builder), the end results (when compared to an unbalanced motor) will be a smoother running motor, with increased power and longer life.

I am speaking here of a *static* balance, rather than a *dynamic* balance. A static balance would consist of matching all reciprocating engine components, by weight, to one another. Although I was unable to find any factory specifications on the static balance of these motors, it is reasonable to assume that the reciprocating assemblies were balanced to within ⅛-ounce of each component. This I'm sure is fine for the average, street driven motor. But if you want the "racer's edge" so to speak, have your components balanced to within ±½ gram; the cost is around $100.

Here is what's involved: The pistons are matched for weight, the lightest one being the target. The remaining pistons are then machined light, to be the same weight as the target piston, with pins installed. The same procedure for the con rods. Piston rings should be all the same weight. The pistons, pins, rings and con rods are then assembled and again matched for weight. If all is in order, the assemblies should be the same ±½gram. The crank throws are then machined and matched for weight of the assembly that will attach to it, again ±½ gram. The end result of all this is a balanced reciprocating assembly that I believe makes a world of difference in engine performance and longevity.

Speed Equipment

A book about the Ford flathead simply would not be a book of any magnitude without some mention of this type of equipment. Hot rodding is as much a part of automotive history as is Henry Ford, Barney Oldfield, Ed Edelbrock or Indianapolis. No other motor has had as much popularity as has the Ford flathead.

Ford publicly announced the new V8 in the *Detroit News* on February 11, 1932 with the first rolling down the assembly line March 9. To make a long story short, the V8 was an overnight success. Strong, light, dependable, affordable and yes, FAST. And just about as fast as Mr. Ford could get these babies off the assembly lines, hot rodders added more carbs and pulled off the stock heads in favor of Offenhauser. This trend continued through the late '30's up to WW II and for the most part was centered in Southern California.

After the war, the word spread to other parts of the country; more flatheads were modified. As more and more Fords were "souped up" the old dirt track racers (hot rodders) found that they had a market for their services and parts. Homemade and shadetree at first, names like Fred Offenhauser, Vic Edelbrock, Phil Weiand, Ed Meyer, Earl Evans, Barney Navarro, Ed Iskenderian, Harman & Collins, Tattersfield & Baron, Knudsen, Windfield, Edmunds, Halibrand, Sharp, Frenzel, Norden, Weber and Smithy, to name but a few, found customers beating a path to their doors. The hot rod industry was born.

The "make a good thing better" trend continued late '54 and and early '55, when the overheads were introduced by Ford and Chevy. But the flathead did not die. It lived. On the streets, at the drags, on the raceways and at Bonneville. It lived through the '50's and early 60's. Slumped a little during the late '60's, caught its breath and fired up again. As we enter the 1980's, I'm here to tell you that the Ford flathead, stock or otherwise, is alive and well in the US of A.

In this book, I have tried to emphasize the availability of stock, flathead motor parts. Speed equipment is no exception. If you're like me, you may be interested. Aluminum heads and manifolds are still available from Offenhauser and Le Baron. Iskenderian still grinds the old ¾ and full-race cams, and Holley makes a super new carb to replace the old Stromberg 97's. Stroker kits, tube headers and ignition parts are also available new. Swap meets always seem to have a good supply of used speed equipment, and the classified sections of the various automotive publications carry their share of

114

used flathead parts. Some of the collector items are a bit more difficult to find, but not impossible. These would include items like Ardun heads, Paxton and McCulloch supercharges. Weiand, Evans and Sharp manifolds are fairly common items and easy to find.

Besides the performance and economy gains (yes Ralph, economy gains), this old time "antique" equipment can really dress up an otherwise ho-hum engine compartment. This old car hobby of ours is a very personal hobby. Tastes and budgets run from McDonalds to epicurean; the beauty of our hobby is truly in the eye of the beholder. A marque club standard, personal preference, time, money and car parts availability are all decisive factors to be reckoned with. If your restoration is not 100 percent original, it's not a sin, only different. The stuff is available, if you want it. It may not be pure or original . . . but it sure is neat.

Conclusion

We've come a long way in our quest to rebuild the Ford Flathead. In the preceding chapters I hope I have covered all the rudimentary ground for your rebuild project.

In this endeavor (or any other for that matter), the key ingredient is PMC; be patient, work methodically and meticulously and above all, keep your cool. The Ford Flathead is a fine machine and will provide you with years of motoring pleasure, once it's back in your car. But it will require some TLC (tender loving care) in the form of some periodic and preventive maintenance. Don't ignore this step and your Ford won't ignore you.

Appendix A
Definition of Fits

The illustrations of fits and tolerances (Appendix B) gives the original clearances established between various parts at the time of manufacture, as well as wear and limit clearances that indicate to what point the clearance may increase before the parts must be replaced. These clearances are based on the parts involved all being at 70 degrees F. The following definitions of the various types of fits are given to assist in arriving at the correct amount of clearance between parts not included in Section B, as well as to give a better appreciation of why the various tolerances must be adhered to. Generally speaking, all bores are made to a standard size (so standard reamers, plug gauges, etc. may be used) with a plus tolerance. The maximum size of the male parts is usually a standard size less the minimum clearance required for the type of fit desired. The minimum size for male parts is the maximum size minus the tolerance.

Wring Fit. A wring fit is the type of fit required between a bore and a plug gauge, when using the plug gauge, to determine the inside diameter of the bore. With a wring fit, it is necessary to turn or wring the plug gauge or part to force it through the bore. This type of fit does not provide space for a film of oil.

Slip Fit. A slip fit exists when the male part is slightly smaller than the female part and involves less clearance than a running fit (below). An example of the minimum allowable clearance for a slip fit would be a piston pin, that, from its own weight, would pass slowly through the connecting rod bushing (with bushing and pin both in a vertical position). In most cases (except where only a limited movement of the parts is involved), slip fits are specified when, due to anticipated expansion of the female part, enough additional clearance will result to change this type of fit to a running fit and provide adequate clearance for a film of oil.

Running Fit. A running fit provides enough clearance for a continuous film of oil between the two parts. A running fit usually requires 0.001-inch for oil film plus a minimum of 0.001-inch for each inch of diameter.

Press Fit. A press fit is one that requires force to enter the male part into the bore. Accepted practice for press fits is to have the male part larger by 0.001 inch for each inch of diameter than the bore into which it is to be pressed.

Shrink Fit. Generally speaking, a shrink fit is tighter than a press fit. The amount of the shrink ranging from 0.001 inch to 0.002 inch for each inch of diameter and in some cases even more. The parts having a press fit may be assembled either by force or by the shrink method. There are two methods of shrinking two parts together, either one of which may be used (both may be used in some instances). One method involves expansion of the female member by heating. The other method involves contracting the male member by chilling with dry ice.

Effect of Expansion on Fits. Allowances are made in establishing fits on parts that are exposed to higher temperature in order to provide for the anticipated expansion of the part during operation and still provide adequate clearance for the type of fit required. Allowances must also be made for unequal expansion of dissimilar materials. Absolute minimum allowance for expansion of parts exposed to flame or exhaust gases (pistons, piston rings, and valves) is 0.001 inch for each inch of diameter or length. In anticipating the expansion of a piston to make allowances for the additional clearance required in the cylinder, 0.001 inch for each inch of diameter is added. In anticipating the expansion of a piston ring, to make allowances for the additional gap required between the ends of the piston ring, 0.001 inch for each linear inch of the part is added.

Appendix B
Guide To Fits

The following is an illustrated guide to measuring the fits and tolerances on your Flathead V8.

Fig. B-1. Measure squareness of bore with top of block, using Vee edge protractor and feeler gauge.

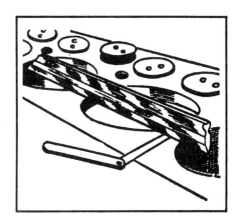

Fig. B-2. Measure flatness of top of block with steel straightedge and feeler gauge.

Fig. B-3. Measure clearance of piston in cylinder with feeler gauge (clearance per inch of piston diameter).

Fig. B-4. Measure piston pin boss bore for parallelism with head, using surface plate and dial gauge.

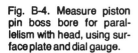

Fig. B-5. Measure roundess of cylinder bore with dial gauge.

Fig. B-6. Measure cylinder bore for taper with inside micrometer.

Fig. B-7. Measure variation in compression height with surface plate and dial gauge.

Fig. B-8. Measure difference in weight between pistons with balance scale.

Fig. B-9. Measure connecting rod bearing for parallelism with piston pin, using aligning fixture and dial gauge.

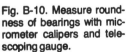

Fig. B-10. Measure roundness of bearings with micrometer calipers and telescoping gauge.

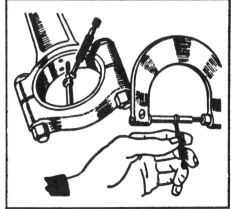

Fig. B-11. Measure bearing to shaft clearance with micrometer calipers and telescoping gauge.

Fig. B-12. Measure end clearance of connecting rod bearing with feeler gauge.

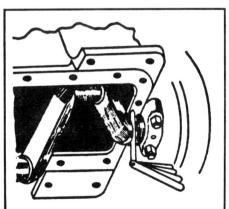

Fig. B-13. Measure end clearance of crankshaft with feeler gauge.

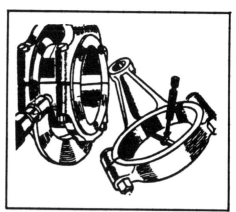

Fig. B-14. Measure clearance of forked-type outside bearing with micrometer calipers and telescoping gauge.

Fig. B-15. Measure straightness of crankshaft with dial gauge.

Fig. B-16. Measure crankpin for taper and roundness with micrometer calipers.

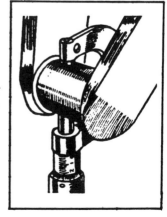

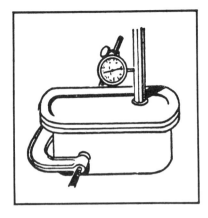

Fig. B-17. Measure clearance between oil pump shaft and bushings with dial gauge.

Fig. B-18. Measure clearance between oil pump housing cover and face of gears with feeler gauge.

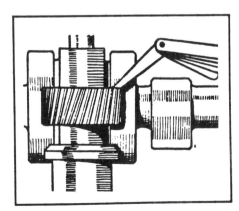

Fig. B-19. Measure clearance between oil pump drive gear and camshaft gear teeth with feeler gauge.

Fig. B-20. Measure clearance between oil pump gear teeth and housing with feeler gauge.

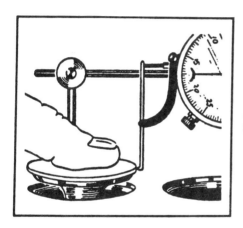

Fig. B-21. Measure clearance between valve stem and valve guide with dial gauge.

Fig. B-22. Measure clearance between valve lifter and lifter guide with micrometer calipers and telescoping gauge.

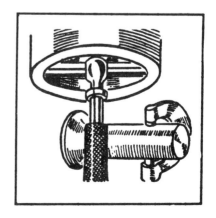

125

Fig. B-23. Measure clearance between rocker arm shaft and bushings with micrometer calipers and telescoping gauge.

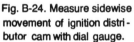Fig. B-24. Measure sidewise movement of ignition distributor cam with dial gauge.

Fig. B-25. Measure camshaft bearing to journal clearance with micrometer calipers and telescoping gauge.

Fig. B-26. Measure straightness of camshaft with dial gauge.

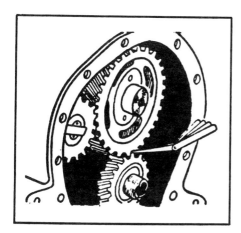

Fig. B-27. Measure clearance between teeth on camshaft and crankshaft timing gears with feeler gauge.

Fig. B-28. Measure lateral trueness of camshaft timing gear with dial gauge.

127

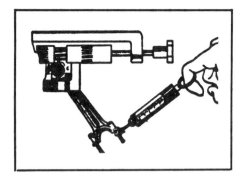

Fig. B-29. Measure fit of floating pin in alloy piston with spring scale.

Fig. B-30. Measure fit of pin using bronze bushings with spring scale.

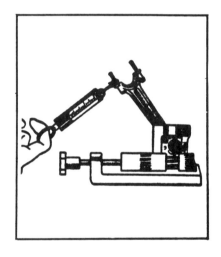

Fig. B-31. Measure clearance of piston rings in grooves with feeler gauge.

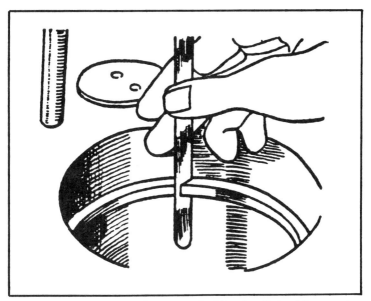

Fig. B-32. Measure clearance between ends of rings in cylinders (clearance per inch of piston diameter).

Appendix C
Torque

Torque is nothing more than a twisting action at a given point. Without some way of measuring torque, it is next to impossible to tighten a series of fasteners to an exact, predetermined tension. The correct sequence and amount of torque is very critical. Applying too little or too much torque will cause leaks, distortion and warpage, which could (and often does) damage or ruin parts.

To properly torque a series of fasteners, use a specific sequence (Fig. C-1). This will apply even pressure on the mating surfaces of the parts being torqued. *Never* torgue two mating

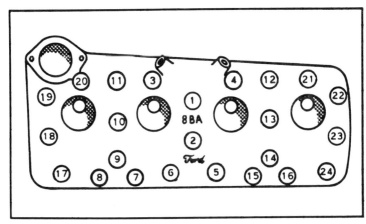

Fig. C-1. Cylinder head bolt tightening sequence.

Table C-1. Torque Wrench Readings.

Main bearing nuts	75 to 80 ft lbs
Connecting rod castellated nuts	35 to 40 ft lbs
Connecting rod self-locking nuts	40 to 45 ft lbs
Cylinder head nuts (cast iron heads)	50 to 60 ft lbs
Cylinder head nuts (aluminum heads)	35 to 40 ft lbs
Flywheel cap screws	65 to 70 ft lbs
Spark plugs (Cast iron heads)	24 to 28 ft lbs
Oil pan (steel)	15 to 18 ft lbs

surfaces by following the fastener pattern clockwise or counterclockwise. Use the sequence shown or one close to it, to be sure of the correct torqueing sequence.

Once your sequence pattern has been determined, snug all the fasteners, then bring your torque up to the desired value in stages of 25 percent, 50 percent, 75 percent and 100 percent of full torque. For example, this section c gives a list of the critical torque values on the flathead motor. The desired torque value for main bearing nuts is 80 ft.-lbs.; 25 percent of this would be 20 ft.-lbs.; 50 percent would be 40 ft.-lbs.; 75 percent would be 60 ft.-lbs. and finally 100 percent would be the correct 80 ft.-lbs.

You should use a torque wrench calibrated in foot-pounds. However, an inch-pound torque wrench will suffice; merely convert inch pounds to foot pounds as follows:

☐ foot pounds x 12 = inch pounds.
☐ inch pounds ÷ 12 = foot pounds.

Appendix D
Specification Tables

The V/8-60 is an almost identical dwarf in appearance, when compared to the larger V/8's. The 60 type is about ⅓ smaller and get its designation from its 60 horsepower rating. These motors were used by Ford from 1937-1940 in addition to the V/8-85. At one time, they were very popular to use in small race boats (cracker box) and midget race cars.

If your car has a V/8-60 and you plan on driving your car a lot, I strongly recommend discarding it and replacing it with a V/8-85. The larger CID & HP output will be to your advantage in the long run. A little more power and a little more torque will be much easier to live with on long hauls. You can easily cruise these motors at 70 mph (if it were legal) and maintain better gas mileage, than trying to push a 60 type to its 45-50 mph limit. The transmission and some under-the-hood hardware will have to be changed, but it's an easy conversion that requires no fabrication modifications to your car. With the proper year V/8-85, your car will remain "pure" with no points loss. Think about it!

Table D-1. Ford V8 60.

Year	Model	Bore	Stroke	C.I.D.	H.P. R.P.M.	Compression ratio
1937	74	2.6	3.2	136	60 @ 3600	6.6 to 1
1938	82A	2.6	3.2	136	60 @ 3600	6.6 to 1
1939	9222A	2.6	3.2	136	60 @ 3600	6.6 to 1
1940	022A	2.6	3.2	136	60 @ 3600	6.6 to 1

Table D-2. Ford V8 and Type 85.

Year	Model	Bore	Stroke	C.I.D.	H.P.	R.P.M.	Compression ratio
1932	18	3.0625"	3.750"	221	65 @	3400	5.5 to 1
1933	40	3.0625"	3.750"	221	75 @	3800	6.33 to 1
1934	40A	3.0625"	3.750"	221	85 @	3800	6.33 to 1
1935	48	3.0625"	3.750"	221	85 @	3800	6.33 to 1
1936	68	3.0625"	3.750"	221	85 @	3800	6.33 to 1
1937	78	3.0625"	3.750"	221		A*	B**
1938	81A	3.0625"	3.750"	221	85 @	3800	6.12 to 1
1939	91A	63.0625"	3.750"	221	85 @	3800	6.15 to 1
1940	01A	3.0626"	3.750"	221	85 @	3800	6.15 to 1
1941	11A	3.0625"	3.750"	221	85 @	3800	6.15 to 1
1942	21A	3.0625"	3.750"	221	90 @	3800	6.2 to 1
1946	69A	3.1875"	3.750"	239.4	100 @	3800	6.75 to 1
1947	79A	3.1875"	3.750"	239.4	100 @	3800	6.75 to 1
1048	89A	3.1875"	3.750"	239.4	100 @	3800	6.75 to 1
1949	98BA	3.875"	3.750"	239.4	100 @	3800	6.8 to 1
1950 to 1953	8BA	3.1875"	3.750"	238.4	100 @	3800	6.8 to 1

A* With aluminum heads. 85   3800
   With cast iron heads. 94   3800

B** With aluminum heads. 6.20 to 1
   With cast iron heads. 7.50 to 1

Table D-3. Mercury V8.

Year	Model	Bore	Stroke	C.I.D.	H.P.	R.P.M.	Compression ratio
1939	99A	3.1875"	3.750"	239.4	95 @	3600	6.3 to 1
1940	09A	3.1875"	3.750"	239.4	95 @	3600	6.3 to 1
1941	19A	3.1875"	3.750"	239.4	95 @	3600	6.3 to 1
1942	29A	3.1875"	3.750"	239.4	100 @	3800	6.4 to 1
1946 to 1948	59A	3.1875"	3.750"	239.4	100 @	3800	6.8 to 1
1949 to 1953	CM	3.1875"	4.000"	255.4	112 @	3800	6.8 to 1

Table D-4. Generator Control Specifications.

PART NUMBER	TYPE	CUT IN VOLTAGE Min.	Max.	VOLTAGE REGULATION at 70 F. Min.	Max.	AMPERAGE REGULATION at 70 F. Min.	Max.
B-10505	Cutout only	6.1	6.3	-	-	-	-
01A-10505	Standasd	6.1	6.3	7.0	7.2	30	33
*68-10505	Two rate relay	6.1	6.3	8.0	8.3	*	*
59A-10505	Standard	6.1	6.3	7.0	7.3	30	33
5EH-10505	Standard	6.1	6.3	7.0	7.3	39	42

*Reduces charging rate when voltage becomes excessive

(3 brush generator only).

Appendix E

Data Tables

Table E-I. Distributor Tune Up Data.

Year	Engine	Spark plug gap	Breaker point gap	Cam angle in degrees	Firing order	Compression pressure at cranking speed	Breaker spring tension
1932-36	V8/85	.025"	.013"	35	15486372	105 psi	22-27 oz.
1937-38	V8/85	.025"	.015"	36	15486372	(95 psi on '32 motors)	
1939-42	V8/85	.025"	.015"	36	15486372	100 psi	20-24 oz.
1946-48	V8/100	.025"	.015"	36	15846372	100 psi	20-24 oz.
1949-53	V8/100	.030"	.015"	29	15846372	100 psi	20-24 oz.
						92 psi	Not applicable

Table E-2. Valve Timing Data.

Year	Valve lift inches	Intake opens B.T.C.	Intake closes A.B.C.	Exhaust opens B.B.C.	Exhaust closes A.T.C.
1932-36	.292	9.5°	54.5°	57.5°	6.5°
1937-53	*.292	0.0°	44.0°	48.0°	6.0°

*1949-53 Intake increased to .296"

Table E-3. Valves and Valve Springs.

Year	Valve Tappet Clearance	Valve Stem Diameter	Valve Stem Clearance in Guide	Valve Springs		
				Pounds of Test		Length in Inches
1932-42	.011"-.013" Intake .011"-.013" Exhaust	.3115"	.0015"-.0035"	36-40	@	2.125"
1946-48	.010"-.012" Intake .014"-.016" Exhaust	.3115"	.0015"-.0035"	36-40	@	2.125"
1943-53	.010"-.012" Intake .014"-.016" Exhaust	.3410"	.0015"-.0035"	36-40	@	2.125"

Table E-4. Pistons, Pins and Rings.

Year	When checking with spring scale		*Piston ring gap		Ring side clearance		Piston pin diameter
	Feeler thickness	Pull on scale	Compression rings	Oil control rings	Compression rings	Oil control rings	
1932-42	A	6-10 lbs	.012	.012	.0015-.003	.0015-.003	.7501-.7504
1946-48	A	6-12 lbs	.012-.017	0.12-.017	.0015-.003	.0015-.003	.7501-.7504
1949-53	.0015	6-12 lbs	.010-.017	.010-.017	.0015-.003	.0015-.003	.7501-.7504

*Fit rings in tapered bores for minimum clearance in tightest portion of ring travel.
A-With cylinder sleeves and aluminum pistons .0025". Without sleeves .002".

Table E-5. Con Rods, Bearings and Crankshaft.

Year	CONNECTING RODS				CRANKSHAFT		
	Crankpin journal diameter	Crankpin to rod clearance	Con rod side play	Con rod bore diameter	Main bearing journal diameter	Main bearing clearance	Crankshaft end thrust
1932-36	1.998-1.999	A	.006-.014	2.220	1.998-1.99	.001-.0025	.002-.006
1937-38	1.998-1.999	A	.006-0.14	2.220	2.398-2.399	.005-.003	.002-.006
1939-42	1.998-1.999	A	.006-0.14	2.220	2.498-2.499	.005-.003	.002-.006
1946-53	2.138-2.139	A	.006-.014	2.298	2.498-2.499	.005-.003	.002-.006

A Cannot be checked in assembly except 1949-53 models. Should be .001"clearance between rod bore and bearing back and additional .001" between inner face of bearing and crankpin.

Table E-6. Cubic Inch Displacement Chart.

Bore in inches	3 ¾"stroke CID in inches	4" stroke CID in inches	4 ⅛" stroke CID in inches
3 1/16	220	236	243
3 3/16	239	255	263
3 5/16	258	274	284
3 5/16+.030	263	281	289
3 ⅜	268	286	296

Table E-7. Fractional Inches Converted to Decimal Inches and Millimeters.

Fraction of Inch	Decimal of Inch	Decimal Millimeters	Fraction of Inch	Decimal of Inch	Decimal Millimeters
1/64	.015625	0.39688	33/64	.515625	13.09690
1/32	.03125	0.79375	17/32	.53125	13.49378
3/64	.046875	1.19063	35/64	.546875	13.89065
1/16	.0625	1.58750	9/16	.5625	14.28753
5/64	.078125	1.98438	37/64	.578125	14.68440
3/32	.09375	2.38125	19/32	.59375	15.08128
7/64	.109375	2.77813	39/64	.609375	15.47816
⅛	.125	3.17501	⅝	.625	15.87503
9/64	.140625	3.57188	41/64	.640625	16.27191
5/32	.15625	3.96876	21/32	.65625	16.66878
11/64	.171875	4.36563	43/64	.671875	17.06566
3/16	.1875	4.76251	11/16	.6875	17.46253
13/64	.203125	5.15939	45/64	.703125	17.85941
7/32	.21875	5.55626	23/32	.71875	18.25629
15/64	.234375	5.95314	47/64	.734375	18.65316
¼	.25	6.35001	¾	.75	19.05004
17/64	.265625	6.74689	49/64	.765625	19.44691
9/32	.28125	7.14376	25/32	.78125	19.84379
19/64	.296875	7.54064	51/64	.796875	20.24067
5/16	.3125	7.93752	13/16	.8125	20.63754
21/64	.328125	8.33439	53/64	.828125	21.03442
11/32	.34375	8.73127	27/32	.84375	21.43219
23/64	.359375	9.12814	55/64	.859375	21.82817
⅜	.375	9.52502	⅞	.875	22.22504
25/64	.390625	9.92189	57/64	.890625	22.62192
13/32	.40625	10.31877	29/32	.90625	23.01880
27/64	.421875	10.71565	59/64	.921875	23.41567
7/16	.4375	11.11252	15/16	.9375	23.81255
29/64	.453125	11.50940	61/64	.953125	24.20942
15/32	.46875	11.90627	31/32	.96875	24.60630
31/64	.484375	12.30315	63/64	.984375	25.00318
½	.5	12.70003	1	1.	25.40005

Index